The Enigmatic Photon

Fundamental Theories of Physics

An International Book Series on The Fundamental Theories of Physics:
Their Clarification, Development and Application

Editor: ALWYN VAN DER MERWE
              University of Denver, U.S.A.

Editorial Advisory Board:

ASIM BARUT, *University of Colorado, U.S.A.*
BRIAN D. JOSEPHSON, *University of Cambridge, U.K.*
CLIVE KILMISTER, *University of London, U.K.*
GÜNTER LUDWIG, *Philipps-Universität, Marburg, Germany*
NATHAN ROSEN, *Israel Institute of Technology, Israel*
MENDEL SACHS, *State University of New York at Buffalo, U.S.A.*
ABDUS SALAM, *International Centre for Theoretical Physics, Trieste, Italy*
HANS-JÜRGEN TREDER, *Zentralinstitut für Astrophysik der Akademie der*
    Wissenschaften, Germany

The Enigmatic Photon

Volume 2: Non-Abelian Electrodynamics

by

Myron Evans
*Department of Physics,
University of North Carolina at Charlotte,
Charlotte, North Carolina, U.S.A.*

and

Jean-Pierre Vigier
*Department of Physics,
Université Pierre et Marie Curie,
Paris, France*

KLUWER ACADEMIC PUBLISHERS
DORDRECHT / BOSTON / LONDON

A C.I.P. Catalogue record for this book is available from the Library of Congress

ISBN 0-7923-3288-1

Published by Kluwer Academic Publishers,
P.O. Box 17, 3300 AA Dordrecht, The Netherlands.

Kluwer Academic Publishers incorporates
the publishing programmes of
D. Reidel, Martinus Nijhoff, Dr W. Junk and MTP Press.

Sold and distributed in the U.S.A. and Canada
by Kluwer Academic Publishers,
101 Philip Drive, Norwell, MA 02061, U.S.A.

In all other countries, sold and distributed
by Kluwer Academic Publishers Group,
P.O. Box 322, 3300 AH Dordrecht, The Netherlands.

Printed on acid-free paper

All Rights Reserved
© 1995 Kluwer Academic Publishers
No part of the material protected by this copyright notice may be reproduced or
utilized in any form or by any means, electronic or mechanical,
including photocopying, recording or by any information storage and
retrieval system, without written permission from the copyright owner.

Printed in the Netherlands

Contents

PREFACE ix

1. $B^{(3)}$ AND THE DIRAC EQUATION 1

1.1 Origins of the Dirac Equation of Motion 2
    1.1.1 Relations between Spin Components 6
1.2 Geometrical Basis [16] of the Dirac Equation 9
    1.2.1 Spinors of the SU(2) Group 9
    1.2.2 Spinors of the SL(2,C) Group 16
1.3 The Free Particle Dirac Equation 19
    1.3.1 Probability Current and Density from the Dirac Equation 23
    1.3.2 Energy Eigenvalues of the Dirac Equation 25
    1.3.3 Standard Representation of the Dirac Equation 26
1.4 The Dirac Equation of e in A_μ: Proof of $B^{(3)}$ from First Principles 28
    1.4.1 Emergence of $B^{(3)}$ from Equation (128) 30
    1.4.2 Complex A_μ: Second Order Process 33
    1.4.3 Complex A_μ: First Order Process 36
1.5 Comparison with the Classical Equation of Motion of e in A_μ 37

2. $B^{(3)}$ AND THE HIGGS PHENOMENON 41

2.1 Cyclically Symmetric Equations for Finite Photon Mass 44
2.2 Link with the Higgs Phenomenon 47
    2.2.1 Spontaneous Symmetry Breaking 55
    2.2.2 $B^{(3)}$ as a Vortex Line in the Vacuum 61

3. $B^{(3)}$ AND NON-ABELIAN GAUGE GEOMETRY 65

    3.1 General Geometrical Theory of Gauge Fields   68
          3.1.1 The Quantization of Charge   74
          3.1.2 Electric and Magnetic Fields in $G_{\mu\nu}$ 75

4. THE O(3) MAXWELL EQUATIONS IN THE VACUUM 79

    4.1 The O(3) Inhomogeneous Maxwell Equations in the Vacuum   79
    4.2 The O(3) Homogeneous Maxwell Equations in the Vacuum   82
    4.3 The Duality Transformation and the O(3) Maxwell Equations   84
          4.3.1 The Dual of $B^{(3)}$ in O(3) Electrodynamics   85
    4.4 Renormalization of O(3) QED   89
    4.5 Isospin and Gauge Symmetry   92

5. $B^{(3)}$ IN UNIFIED FIELD THEORY 95

    5.1 Summary of the Non-abelian Features of $W_{3\mu}^{(1)}$ and $X_\mu^{(1)}$ 98
    5.2 Specific Effects of $B^{(3)}$ in GWS Theory   103
    5.3 SSB and Photon Mass in GWS   107
          5.3.1 SSB as the Source of Photon Mass in Abelian Theory   109
          5.3.2 SSB as the Source of Photon Mass in Non-Abelian theory   111

6. $B^{(3)}$ IN QUANTUM ELECTRODYNAMICS 113

    6.1 Canonical Quantization and $B^{(3)}$ 113
    6.2 The Effect of $B^{(3)}$ on Renormalizability in QED   116
    6.3 $B^{(3)}$ and the Electron's Magnetic Moment   118
          6.3.1 Calculation of the Anomalous Magnetic Moment of the Electron in QED   119
          6.3.2 Origin of the Convergent Vertex in QED   121

7. SUMMARY OF ARGUMENTS AND SUGGESTIONS FOR EXPERIMENTAL VERIFICATION   123

APPENDICES

A. The O(3) Electromagnetic Field Tensor, $G_{\mu\nu}$, in the Circular Basis (1), (2), (3)   135

B. The O(3) Covariant Derivative (D_μ) in the Basis (1), (2), (3)   139

C. The Structural Analogy between NAE and General Relativity   143

D. Structure of the Field Tensor $G_{\mu\nu}^{(1)}$ of Non-Abelian Electrodynamics   147

E. Some Details of the Non-Abelian Maxwell Equations in the Vacuum   159

REFERENCES   163

INDEX   167

CONTENTS OF VOLUME 1   173

Preface

This second volume of *The Enigmatic Photon* opens with a demonstration of the existence of the longitudinal vacuum field $B^{(3)}$ (Vol. 1) from a consideration of the Dirac equation of one electron, e, in a circularly polarized electromagnetic field, represented by the four-potential A_μ. This results in the key inference that the interaction Hamiltonian formed between the electron's intrinsic angular momentum and the field is governed entirely by $B^{(3)}$. The latter is thereby shown to be a fundamental intrinsic property of vacuum electromagnetism. The second and succeeding chapters develop the role of $B^{(3)}$ in field theory. Chapter two deals with the Higgs phenomenon and spontaneous symmetry breaking of the vacuum. The existence of the longitudinal $B^{(3)}$ implies that there is finite photon mass, which is made compatible in Chap. 2 with gauge invariance of the second kind. In Chaps. 3 and 4, the non-Abelian nature of the relation between the spin field $B^{(3)}$ and the plane waves $B^{(1)}$ and $B^{(2)}$ is used to develop a self-consistent view of vacuum electromagnetism using an O(3) gauge geometry. This leads to the non-Abelian vacuum Maxwell equations in Chap. 4, the technical details of which are relegated to Appendices. The latter provide detailed checks on the self-consistency of the new theory, in which Yang-Mills type isospin indices are identified with circular indices, (1), (2) and (3), of three dimensional space. In Chap. 5, a development is given of the role of $B^{(3)}$ in unified (electroweak) theory, in particular its role in GWS theory. In Chap. 6, the effect of $B^{(3)}$ on quantum electrodynamics is developed, and it is shown that $B^{(3)}$ is consistent with the powerful results of QED, for example the latter's ability to produce the anomalous magnetic moment of the electron with great precision. Finally, in Chap. 7, a summary of the major results of both volumes is given, a summary which shows that the discovery of $B^{(3)}$ is of central importance in contemporary field theory. For example it shows conclusively that the photon, if it is a particle, must have mass, and means that the gauge group of electromagnetism is O(3), and not O(2) = U(1) as thought conventionally. Furthermore, the existence of $\boldsymbol{B}^{(3)}$ in the vacuum can be shown experimentally by using powerful microwave pulses to magnetize an electron plasma set up in an inert gas such as helium. The precise conditions for such an experiment are given in Chap. 7. This technique isolates the field $\boldsymbol{B}^{(3)}$ through its characteristic *square root* power density profile ($I_0^{1/2}$), i.e., the magnetization set up by $\boldsymbol{B}^{(3)}$ in the plasma is proportional to $I_0^{1/2}$. This is not possible if $\boldsymbol{B}^{(3)}$ were zero. The magnetization from the wave fields $\boldsymbol{B}^{(1)}$ and $\boldsymbol{B}^{(2)}$ is to order I_0. Thus $\boldsymbol{B}^{(3)}$ is an experimental observable, and exists in the vacuum, proving that the wave fields $\boldsymbol{B}^{(1)}$ and $\boldsymbol{B}^{(2)}$ and the spin field $\boldsymbol{B}^{(3)}$ are linked through the non-Abelian defining algebra,

$$\boldsymbol{B}^{(1)} \times \boldsymbol{B}^{(2)} = i B^{(0)} \boldsymbol{B}^{(3)*} \text{ et cyclicum,}$$

in the vacuum. The numerous fundamental consequences of this algebra are discussed throughout these two volumes, which view electromagnetism in an entirely original way.

We owe a great debt of gratitude to Dr. Laura J. Evans, whose highly professional camera-ready preparation was of key importance to the whole two volume project.

Many interesting discussion before and during production are acknowledged with several colleagues, including Keith A. Earle, Gareth J. Evans, the late Stanisław Kielich, Mikhail A. Novikov, Mark P. Silverman, Boris Yu Zel'dovich, and others. Last, but by no means least, Professor Alwyn van der Merwe is acknowledged with gratitude for the opportunity of producing these volumes in his prestigious series.

Charlotte, North Carolina, U.S.A. and Myron W. Evans
Craigcefnparc, Wales
Paris, France Jean-Pierre Vigier
July, 1994

Chapter 1. $B^{(3)}$ and the Dirac Equation

In Vol. 1, several methods were used to infer that there exists in free space a spin field $B^{(3)}$ of electromagnetic radiation, which is the recently discovered [1-12] magnetizing field of light. In the opening chapter of this volume, the Dirac equation of motion is used to prove that $B^{(3)}$ emerges directly from the consideration of the action of electromagnetic radiation on one electron. The Dirac equation is a relativistically correct and physically meaningful quantum counterpart of the relation between mass and energy in classical special relativity, and in the non-relativistic limit reduces to a Schrödinger equation of quantum mechanics. The Dirac equation for a free electron indicates that it, the electron, has an intrinsic spin angular momentum, which is essentially a consequence of the geometry of space-time expressed in terms of spinors. This spin angular momentum has eigenvalues $\pm \hbar/2$, has no classical counterpart, and remains non-zero in the non-relativistic limit [13] of the Dirac equation. In consequence, the Dirac equation is able to account for the anomalous Zeeman effect [14] and the results of the Stern-Gerlach experiment [15]. Its major importance is underlined by the fact that it predicts the existence of anti-particles through the concept of the Dirac sea [16], and for these reasons has supplanted the direct quantum equivalent of the Einstein equation, the Klein-Gordon equation of motion [17]. An inference based directly on the Dirac equation is therefore based firmly in fundamental theory. In what follows, the Dirac equation reveals the presence of $B^{(3)}$ in the interaction of electromagnetic radiation with one electron, and so $B^{(3)}$ is an observable and is present in free space, as inferred in several different ways in Vol. 1. The magnetic field of light, $B^{(3)}$, forms an interaction Hamiltonian with the magnetic dipole moment formed from the *intrinsic* electronic spin. The latter does not exist in classical field theory, and so this effect of $B^{(3)}$ occurs in addition to its induction of a classically based *orbital* electronic angular momentum as described in Chap. 12 of Vol. 1.

1.1 ORIGINS OF THE DIRAC EQUATION OF MOTION

The Dirac equation of motion emerged from the attempts to apply the new quantum theory to special relativity. As ably described in many texts [13–17], the direct attempt at quantization of special relativity resulted in negative probability density and energy eigenstates, which proved to be physically uninterpretable. In this section, a brief account is given based on Ryder [16], of the methods used in the derivation of the Klein-Gordon equation for a particle within the framework of special relativity.

From the classical theory of special relativity emerges the following relation,

$$En^2 = m_0^2 c^4 + p^2 c^2, \qquad (1)$$

between energy and momentum. Here m_0 is the rest mass of the particle, c the velocity of light, and p its linear relativistic momentum. In a frame of reference in which the particle has no linear momentum, this equation reduces to Einstein's equation for rest energy,

$$En = m_0 c^2, \qquad (2)$$

showing that mass is energy. The rest mass m_0 is frame invariant, and therefore the rest energy remains the same in all Lorentz frames of reference, a familiar result from special relativity. The Klein-Gordon equation is the result of applying the fundamental quantum mechanical axioms [17] directly to Eq. (1);

$$\hat{p} \rightarrow -i\hbar \nabla, \qquad En \rightarrow i\hbar \frac{\hat{\partial}}{\partial t}. \qquad (3)$$

The wavefunction of the equation, denoted ϕ, is that for a particle with no spin, a scalar particle with only one component, a particle that can also be interpreted as a scalar *field* [16]. The Klein-Gordon equation of motion is therefore an equation of relativistic quantum field theory. Substituting Eq. (3) in Eq. (1) gives

$$\hat{\Box} \phi = -\xi^2 \phi, \qquad (4)$$

where

$$\hat{\Box} := \frac{1}{c^2}\frac{\partial^2}{\partial t^2} - \nabla^2, \quad \xi := \frac{m_0 c}{\hbar}. \tag{5}$$

The Proca equation for a photon with mass, Eq. (9b) of Vol. 1, is Eq. (4) with ϕ replaced by the *gauge* field A_μ. In the non-relativistic approximation, the kinetic energy of a free particle is the familiar expression

$$T = \frac{1}{2}m_0 v^2 = \frac{p^2}{2m_0}, \tag{6}$$

where $p = m_0 v$ is the particle's linear momentum. Applying the fundamental axioms (3) to Eq. (6) gives the Schrödinger equation of motion of a free particle,

$$\frac{\hbar^2}{2m_0}\nabla^2\phi = -i\hbar\frac{\partial\phi}{\partial t}. \tag{7}$$

Equation (7) is the non-relativistic approximation to Eq. (4). The wave function of the two equations is ϕ, and so the conjugate product $\phi^*\phi$ is a probability density in the Born interpretation of quantum mechanics, suggested in 1926 [18],

$$\rho = \phi^*\phi. \tag{8}$$

In special relativity, however, ρ must be the time-like part of a current-density four-vector,

$$j_\mu := (\mathbf{J}, i\rho c), \tag{9}$$

and ρ must be covariant under Lorentz transformation. The space-like component of j_μ is the *probability current*, $\mathbf{J}$, defined by

$$\mathbf{J} = -\frac{i\hbar}{2m_0}(\phi^*\nabla\phi - \phi\nabla\phi^*), \tag{10}$$

which, using Eq. (3), can be written as

$$J = \frac{1}{2m_0}(\phi^*p\phi + \phi p^*\phi^*) = \frac{1}{2m_0}(\phi^*p\phi - \phi p\phi^*), \quad p = -p^*, \quad (11)$$

which has the units of linear velocity. Since $m_0 J$ has the units of linear momentum it must form part of a momentum-energy four-vector in special relativity, as indicated by Eq. (9). Since j_μ is a four-vector it follows that the equation

$$\frac{\partial j_\mu}{\partial x_\mu} = 0 \quad (12)$$

is a *continuity equation* in Minkowski notation. In vector notation

$$\nabla \cdot J - \frac{\partial \rho}{\partial t} = 0, \quad (13)$$

which is a continuity equation for J and ρ. Equation (12) is a conservation theorem which shows that j_μ is conserved, essentially an outcome of Noether's Theorem [16]. Equation (8) must therefore be an approximation to the correctly covariant

$$\rho = -\frac{i\hbar}{2m_0 c^2}\left(\phi^*\frac{\partial \phi}{\partial t} - \phi\frac{\partial \phi^*}{\partial t}\right). \quad (14)$$

A set of equations precisely analogous with Eqs. (9), (10) and (14) can be constructed [18] for electric charge-current density, whose four-vector is $j_\mu^{(e)} = ej_\mu$, where e is the electronic charge, and which appears in the covariant formulation of Maxwell's equations for the interaction of electromagnetic radiation and matter.

With these definitions, Eq. (12) can be expressed as

$$\frac{\partial j_\mu}{\partial x_\mu} = \frac{i\hbar}{2m_0}(\phi^*\Box\phi - \phi\Box\phi^*) = 0, \quad (15)$$

but it is well known [16] that this interpretation collapses because ρ from Eq. (14) can become negative, whereas any probability density must be positive definite to retain physical meaning. The essential mathematical reason for this is that the Klein-Gordon equation is a second order differential equation in which both ρ and $\partial\rho/\partial t$ can be fixed arbitrarily at any given instant in time [16]. The equation cannot therefore be interpreted physically as a single particle wave equation of quantum mechanics with wave function ϕ.

The Dirac equation of motion, on the other hand, is a relativistically correct generalization of the Schrödinger equation which surmounts these difficulties through the use of the appropriate space-time group SL(2,C) [16,17], which is the group which represents the Lorentz transformation with four-spinors. Comparison with experimental data of many different kinds has shown that the Dirac equation is an accurate description of nature at a fundamental level. It is essentially a geometrical relation between different four-spinors, and leads to a positive definite ρ *provided that there exist particles and anti-particles with an intrinsic angular momentum with eigenvalues* $\pm\hbar/2$. These are fermions which obey Fermi-Dirac statistics. This type of angular momentum is essentially a Lorentz transformation property of four-spinors, and is therefore a direct consequence of special relativity itself. It does not imply that fermions spin about an axis fixed in the particle, so that the often used term "spin angular momentum" is slightly misleading. A vector field has an integral intrinsic angular momentum [19], and the spinor field has half integral spin. These intrinsic angular momenta exist regardless of the spatial description of a field, such as an electromagnetic field. The intrinsic integral spin field $\boldsymbol{B}^{(3)}$ of free space electromagnetism corresponds to the intrinsic half integral spin of the electron. The Dirac equation shows that one quantity cannot exist without the other when considering the interaction of e with a classical A_μ. The vector field $\boldsymbol{B}^{(3)}$ has an intrinsic, or built-in, angular momentum [19] of unity, and this, being phase free, has nothing to do with the spatial distribution of the transverse fields $\boldsymbol{B}^{(1)}$ and $\boldsymbol{B}^{(2)}$ (Vol. 1). In the same way, the electron has a built-in half-integral angular momentum, which has nothing to do with the spatial distribution of charge. Both types of angular momenta are relativistically invariant and both are transformation properties. In the non-relativistic limit, they are transformation properties under rotations in space, using vectors of the O(3) group for $\boldsymbol{B}^{(3)}$ and spinors of the SU(2) group for the electron. Thus $\boldsymbol{B}^{(3)}$ is a fundamental and generally applicable outcome of the Dirac equation of e in A_μ and multiplies a magnetic dipole moment formed from the electron's intrinsic spin to form part of the interaction Hamiltonian. *Without this term, the electron's spin could not contribute to the Hamiltonian.* We conclude that given the intrinsic spin of the electron, the Dirac equation describing the interaction of e with A_μ implies the existence of the intrinsic spin field $\boldsymbol{B}^{(3)}$ of free space electromagnetism through an irremovable interaction term in the Hamiltonian.

It is important to note that $\boldsymbol{B}^{(3)}$ in this description is a *classical* field, and can be deduced from the classical description of e in A_μ given in Chap. 12 of Vol. 1. The intrinsic electronic spin has no classical counterpart, and the classical Hamilton-Jacobi description of e in A_μ given in Chap. 12 of Vol. 1 involves only *orbital* electronic angular momentum and an induced magnetic dipole moment. The quantum mechanical Dirac equation produces an intrinsic electronic spin which gives rise to a permanent magnetic dipole moment. The most rigorous treatment of e in A_μ occurs in quantum electrodynamics [16] where the field is quantized.

1.1.1 RELATIONS BETWEEN SPIN COMPONENTS

The free space Proca equation (Vol. 1),

$$\hat{\Box} A_\mu = -\xi^2 A_\mu, \qquad (16)$$

is a physically meaningful wave equation for the photon, regarded as a particle with mass. This is so despite the fact that Eqs. (4) and (16) are identical in structure, and despite the fact that Eq. (4) is not physically meaningful as a wave equation. Particles (bosons) described by a gauge field are therefore fundamentally different in nature from fermions whether or not the bosons have mass. The Dirac equation for a fermion of mass m_0 also has the same structure as Eqs. (4) and (16),

$$\hat{\Box} \psi = -\xi^2 \psi, \qquad (17)$$

but now ψ is a four-spinor, and not a four-vector such as A_μ, or a scalar such as ϕ. However, each (scalar) component of both Eqs. (16) and (17) must obey Eq. (4) by definition, since both A_μ and ψ are made up of scalar components. Whether these equations are physically meaningful or not therefore depends on the nature of the wave-functions ϕ, A_μ, and ψ. The Dirac and Proca equations can *both* be derived [20] by considering the transformation of spinors under the Lorentz group, and are therefore simply relations between

Origins of the Dirac Equation of Motion

spin components in space-time. In the limit of zero mass, the Dirac equation becomes the Weyl equation and the Proca the d'Alembert equation for free space electromagnetism, respectively;

$$\Box \psi = 0, \tag{18a}$$

$$\Box A_\mu = 0. \tag{18b}$$

These equations describe respectively the massless neutrino (a fermion) and the massless photon (a boson). The Klein-Gordon equation of motion, Eq. (4), cannot by definition be a relation between spin components, because it describes the scalar (one component and spinless) wave function ϕ.

The way in which spin components are *related* in space-time determines whether the particle being described is a fermion (half-integral spin) or a boson (integral spin). Since there is one space-time, A_μ and ψ are two different geometrical representations of wave-functions arising from the same four dimensional source. In this chain of reasoning, the probability densities from the Dirac and Proca equations must both be physically acceptable and positive definite, and this is indeed the case. Following Barut [17] for example, the Maxwell equations in matter can be written as

$$\frac{\partial F_{\mu\nu}}{\partial x_\nu} = ej_\mu, \tag{19}$$

and this expression is also an equation of continuity [17], because

$$\frac{\partial}{\partial x_\mu}\left(\frac{\partial F_{\mu\nu}}{\partial x_\nu}\right) = e\frac{\partial j_\mu}{\partial x_\mu} = 0. \tag{20}$$

Equation (20) is the result of the fact that $F_{\mu\nu}$ is an anti-symmetric tensor,

$$F_{\mu\nu} := \frac{\partial A_\nu}{\partial x_\mu} - \frac{\partial A_\mu}{\partial x_\nu}. \tag{21}$$

One of the major advantages of the Dirac equation is that it gives a positive definite probability density, and

8   Chapter 1. $B^{(3)}$ and the Dirac Equation

cures this ailment of the Klein-Gordon equation. This
advantage is illustrated [17] by a field-particle *interaction
equation* such as that describing the electron in the electromagnetic field,

$$\frac{\partial F_{\mu\nu}}{\partial x_\nu} = ej_\mu = ec\bar{\psi}\gamma_\mu\psi. \qquad (22)$$

This equation shows that the expectation value of the quantum mechanical Dirac matrix γ_μ is the Maxwellian four-current. The eigenfunctions used in the evaluation of the expectation value of γ_μ are the Dirac four-spinor ψ and its *adjoint spinor*, $\bar{\psi}$. These concepts will be explained in the following sections. Equation (22), illustrating the interaction of an electron *e* with the electromagnetic field A_μ, shows that the physically meaningful nature of the Dirac and Maxwell equations is based on the presence of *more than one spin component* in the particle or field being described. We shall show that the (Maxwellian) spin field $B^{(3)}$, the newly recognized (Vol. 1) *magnetizing field of light*, is a direct outcome of the Dirac equation for the interaction of an electron, *e*, with A_μ, i.e., a direct result of a fundamental particle-field equation such as Eq. (22). The interaction occurs through a term in the Hamiltonian eigenvalue of the type

$$H^{(int)} = -\frac{e\hbar}{2m_0}\boldsymbol{\sigma}\cdot\boldsymbol{B}^{(3)} := -\frac{e}{m_0}\boldsymbol{S}\cdot\boldsymbol{B}^{(3)}, \qquad (23)$$

where $\boldsymbol{\sigma}$ is a Pauli two-spinor [16] and where $\boldsymbol{S}$ is the spin angular momentum of the electron, with eigenvalues $\pm\hbar/2$. Therefore, the well known fundamental property $\boldsymbol{S}$ of the electron must interact with an electromagnetic field, considered classically, through the novel magnetizing field $B^{(3)}$, a *phase-free* magnetic flux density in free space, *and with $B^{(3)}$ only*.

The magnetizing field of light, $B^{(3)}$, emerges from the Dirac equation of motion of *e* in the field of light in the same way precisely as $\boldsymbol{S}$ emerges. Therefore $B^{(3)}$ is a fundamental property of the classical electromagnetic field (and of the photon) in the same way that $\boldsymbol{S}$ is a fundamental quantum mechanical property of the electron. This conclusion rigorously reinforces the arguments for $B^{(3)}$ presented in Vol. 1 of this book, and shows that both $\boldsymbol{S}$ and $B^{(3)}$ are direct consequences of space-time geometry itself at the most fundamental level in contemporary thought. It is therefore

Origins of the Dirac Equation of Motion

incorrect to assert that $B^{(3)}$ is zero, because this assertion [21] destroys the validity of the Dirac equation itself. Similarly, it is incorrect to assert that S is zero for the electron.

1.2 GEOMETRICAL BASIS [16] OF THE DIRAC EQUATION

1.2.1 SPINORS OF THE SU(2) GROUP

The Dirac equation of motion is a description of the Lorentz boost transformation [16], a description based on a representation of space-time in terms of four-spinors rather than four-vectors. Essentially, spinors, as the name implies, introduce intrinsic spin into the space-time trajectory of a fermion such as an electron or neutrino, which thereby acquires a *helicity* [22]. Furthermore, the fundamentally geometrical nature of spinors allows only two components of spin, for example $\langle S \rangle = \pm\hbar/2$ as in Sec. 1.1. Since four-spinors and four-vectors are both methods of describing space-time, there must have been an experimental basis for the choice of spinors by Dirac [23]. This basis is well known to have included the Stern-Gerlach experiment [24], in which a beam of silver atoms travelling through an inhomogeneous magnetic field is split into *two components, and two only*; and the failure of the Klein-Gordon equation of motion as described in Sec. 1.1.

Adapting the description by Ryder [16], the simplest geometrical properties of spinors can be constructed from the SU(2) group of unitary matrices with complex coefficients [25, 26]. The main purpose of this volume is to derive in various ways the rigorous basis for the novel $B^{(3)}$ field, and therefore we restrict our development of the theory of spinors to the minimum necessary for comprehension.

The SU(2) spinor is denoted by a two component column vector with complex coefficients,

$$\xi = \begin{pmatrix} \xi_1 \\ \xi_2 \end{pmatrix}, \qquad (24)$$

where the *complex* ξ_1 and ξ_2 are related to the real X, Y, and Z Cartesian components of a three-vector of the rotation group O(3) of Vol. 1 by

$$X = \frac{(\xi_2^2 - \xi_1^2)}{2}, \qquad (25a)$$

$$Y = \frac{(\xi_1^2 + \xi_2^2)}{2i}, \tag{25b}$$

$$Z = \xi_1 \xi_2. \tag{25c}$$

Therefore

$$|\mathbf{R}| = (X^2 + Y^2 + Z^2)^{\frac{1}{2}} = (\xi_1^2 \xi_2^2)^{\frac{1}{2}}, \tag{26}$$

which shows that the radius vector $\mathbf{R}$ is represented by two spinor components or coordinates, ξ_1 and ξ_2, as opposed to three Cartesian coordinates. It may then be shown [16] that an SU(2) transformation on ξ is equivalent to an O(3) transformation on the column vector $\begin{pmatrix} X \\ Y \\ Z \end{pmatrix}$. A *rotation* of the $\mathbf{R}$ vector in O(3) can be represented by a unitary transformation matrix, whose inverse is equal to its transpose [25]. Similarly, the rotation of the spinor in SU(2) can be represented by a complex unitary matrix, U, whose determinant is unity,

$$U = \begin{pmatrix} a & b \\ -b^* & a^* \end{pmatrix}, \quad |a|^2 + |b|^2 = 1, \quad UU^+ = UU^{-1} = 1. \tag{27}$$

Here the superscript + denotes *Hermitian transpose*, which is the transpose accompanied by complex conjugation of each matrix element. The superscript -1 denotes the Hermitian inverse, which is the inverse of the matrix accompanied by complex conjugation of each element. The superscript * denotes *complex conjugate*, and the superscript T denotes the transpose without complex conjugation. The SU(2) group is the group of these 2 x 2 unitary, complex matrices [16]. A transformation of the spinor ξ in SU(2) is equivalent to a transformation of the three-vector $\mathbf{R}$ in O(3), and is given by

$$\xi \to U\xi, \quad \xi^+ \to \xi^+ U^+, \tag{28}$$

where [16]

$$U^+ = \begin{pmatrix} a^* & -b \\ b^* & a \end{pmatrix} = U^{-1}. \tag{29}$$

Geometrical Basis [16] of the Dirac Equation

The inner (or scalar) product of two spinors,

$$\xi^+\xi := \begin{pmatrix} \xi_1^* & \xi_2^* \end{pmatrix} \begin{pmatrix} \xi_1 \\ \xi_2 \end{pmatrix} = \xi_1^*\xi_1 + \xi_2^*\xi_2, \qquad (30)$$

is frame invariant, and the outer product,

$$\xi\xi^+ := \begin{pmatrix} \xi_1 \\ \xi_2 \end{pmatrix}\begin{pmatrix} \xi_1^* & \xi_2^* \end{pmatrix} = \begin{pmatrix} \xi_1\xi_1^* & \xi_1\xi_2^* \\ \xi_2\xi_1^* & \xi_2\xi_2^* \end{pmatrix}, \qquad (31)$$

is a Hermitian matrix, i.e., is a square matrix which is unchanged by taking the transpose of its complex conjugate. Under a SU(2) transformation, it becomes, from Eqs. (28),

$$\xi\xi^+ \to U\xi\xi^+ U^+. \qquad (32)$$

In order to introduce *the Pauli spinors* σ_X, σ_Y, and σ_Z, note, following Ryder [16] that

$$\zeta\xi^* := \begin{pmatrix} 0 & -1 \\ 1 & 0 \end{pmatrix}\begin{pmatrix} \xi_1^* \\ \xi_2^* \end{pmatrix} = \begin{pmatrix} -\xi_2^* \\ \xi_1^* \end{pmatrix} \qquad (33)$$

transforms in the same way as ξ under SU(2). This property is denoted by

$$\xi \sim \zeta\xi^*. \qquad (34)$$

Similarly, taking the complex conjugates on both sides

$$\xi^* \sim \zeta\xi, \qquad (35)$$

and

$$\xi^+ := (\xi^*)^T \sim (\zeta\xi)^T = (-\xi_2 \ \xi_1). \qquad (36)$$

Therefore

$$\xi\xi^+ \sim \begin{pmatrix} \xi_1 \\ \xi_2 \end{pmatrix}(-\xi_2 \ \xi_1) := -H = \begin{pmatrix} -\xi_1\xi_2 & \xi_1^2 \\ -\xi_2^2 & \xi_1\xi_2 \end{pmatrix}. \tag{37}$$

However, we know that

$$\xi\xi^+ \to U\xi\xi^+ U^+ \tag{38}$$

so

$$H \to UHU^+ = UHU^{-1} \tag{39}$$

under SU(2). Using the notation

$$h := \boldsymbol{\sigma} \cdot \boldsymbol{r} = \sigma_X X + \sigma_Y Y + \sigma_Z Z = \begin{pmatrix} 0 & 1 \\ 1 & 0 \end{pmatrix} X + \begin{pmatrix} 0 & -i \\ i & 0 \end{pmatrix} Y$$
$$+ \begin{pmatrix} 1 & 0 \\ 0 & -1 \end{pmatrix} Z = \begin{pmatrix} Z & X-iY \\ X+iY & -Z \end{pmatrix}, \tag{40}$$

it is found that h transforms under SU(2) in the same way as H,

$$h \to UhU^+ := h', \tag{41}$$

and that this is equivalent to a *rotation* in O(3) of the three-vector

$$\boldsymbol{R} = X\boldsymbol{i} + Y\boldsymbol{j} + Z\boldsymbol{k}. \tag{42}$$

The transformation $h \to h'$ under SU(2) takes place via the Pauli spinors σ_X, σ_Y, σ_Z in such a way that

$$det \ h = det \ h', \tag{43}$$

i.e.,

$$X^2 + Y^2 + Z^2 = X'^2 + Y'^2 + Z'^2 \tag{44}$$

as required for a rotation of $\boldsymbol{R}$ in O(3). Furthermore, if U is unitary, i.e., if $U^+ = U^{-1}$, the transformation $h \to UhU^+ = h'$

Geometrical Basis [16] of the Dirac Equation

means that both h and h' are Hermitian and traceless.
It is therefore possible to identify the elements of H and h,

$$\begin{pmatrix} \xi_1\xi_2 & -\xi_1^2 \\ \xi_2^2 & -\xi_1\xi_2 \end{pmatrix} = \begin{pmatrix} Z & X-iY \\ X+iY & -Z \end{pmatrix}, \quad (45)$$

leading to Eqs. (25). Note that a factor 1/2 appears on the right hand sides of Eqs. (25a) and (25b), *and this is the origin of the half integral spin of the fermion, a purely geometrical consequence of $H = h$ in Eq. (45)*. This point can be clarified [16] by using the fact that under SU(2)

$$\xi \rightarrow \begin{pmatrix} a & b \\ -b^* & a^* \end{pmatrix}\begin{pmatrix} \xi_1 \\ \xi_2 \end{pmatrix} = \begin{pmatrix} a\xi_1 + b\xi_2 \\ -b^*\xi_1 + a^*\xi_2 \end{pmatrix}, \quad (46)$$

so

$$\xi_1 \rightarrow a\xi_1 + b\xi_2 = \xi_1', \quad \xi_2 \rightarrow -b^*\xi_1 + a^*\xi_2 = \xi_2'. \quad (47)$$

Using Eqs. (25) in the transformed frame, e.g.,

$$Z' = \xi_1'\xi_2' \quad (48)$$

and similarly for X' and Y'; and eliminating ξ_1, ξ_2, ξ_1', ξ_2' between the equations, it can be shown that $\xi \rightarrow \xi'$ is equivalent to

$$X' = X\cos\alpha + Y\sin\alpha, \quad Y' = -X\sin\alpha + Y\cos\alpha,$$
$$Z' = Z, \quad (49)$$

provided that $a = e^{i\alpha/2}$, $b = 0$. In O(3), Eq. (49) is a rotation through an angle α about the Z axis [16]. Therefore the transformation $\xi \rightarrow \xi'$ in SU(2) is entirely equivalent to a rotation in O(3). This can be so if and only if a factor 1/2 appears in the exponent defining a, and this factor 1/2 carries through to describe the intrinsic spin of a fermion. To understand the origin of fermions it is therefore necessary to understand the theory of spinors in SU(2) and, as described in the following section, SL(2,C).

The unitary matrix U of SU(2) is therefore

$$U = \begin{pmatrix} a & b \\ -b^* & a^* \end{pmatrix} = \begin{pmatrix} e^{i\alpha/2} & 0 \\ 0 & e^{-i\alpha/2} \end{pmatrix}, \tag{50}$$

and is equivalent to, i.e., isomorphic with, the O(3) rotation matrix [16],

$$R = \begin{pmatrix} \cos\alpha & \sin\alpha & 0 \\ -\sin\alpha & \cos\alpha & 0 \\ 0 & 0 & 1 \end{pmatrix}, \tag{51}$$

which is deduced from trigonometry in free space. The isomorphism between U and R is denoted by

$$U \leftrightarrow R, \tag{52}$$

i.e., the *complex* unitary rotation matrix U of SU(2) is isomorphic with the *real* unitary rotation matrix R of O(3). The two matrices describe a rotation in free space, one in terms of two component spinors, the other in terms of three component vectors.

The equivalent of the infinitesimal rotation generators of O(3) (see Vol. 1) can also be defined from U as follows:

$$\frac{\sigma_z}{2} = \frac{1}{i}\frac{dU}{d\alpha}\bigg|_{\alpha=0} = \frac{1}{2}\begin{pmatrix} 1 & 0 \\ 0 & -1 \end{pmatrix}, \tag{53}$$

showing that the Pauli matrix $\frac{\sigma_z}{2}$ is an infinitesimal rotation generator of SU(2), isomorphic with J_z of O(3). Using a formal Maclaurin expansion for the infinitesimal angles $\delta\alpha$,

$$U(\delta\alpha) = 1 + \frac{i\sigma_z}{2}\delta\alpha + \ldots, \tag{54}$$

from which

Geometrical Basis [16] of the Dirac Equation

$$e^{i\sigma_z\alpha/2} = 1 + i\sigma_z\frac{\alpha}{2} + \ldots := \begin{pmatrix} 1 & 0 \\ 0 & 1 \end{pmatrix}\cos\frac{\alpha}{2} + i\begin{pmatrix} 1 & 0 \\ 0 & -1 \end{pmatrix}\sin\frac{\alpha}{2} \quad (55)$$

$$= \cos\frac{\alpha}{2} + i\sigma_z\sin\frac{\alpha}{2}$$

to first order in $\alpha/2$. Generalizing finally, to a finite rotation through θ about an axis $\mathbf{n}$ [16], we obtain, with $\boldsymbol{\theta} := \theta\mathbf{n}$,

$$e^{i\boldsymbol{\sigma}\cdot\boldsymbol{\theta}/2} = e^{i\boldsymbol{\sigma}\cdot\mathbf{n}\theta/2} = \cos\frac{\theta}{2} + i\boldsymbol{\sigma}\cdot\mathbf{n}\sin\frac{\theta}{2}. \quad (56)$$

This is the basic structure in space that gives rise to the Dirac equation in the SL(2,C) group of space-time. For a rotation about Z, for example, $\boldsymbol{\sigma}\cdot\mathbf{n} = \sigma_z$ and Eq. (55) is recovered. This derivation has emphasized that the Dirac equation is a direct consequence of the geometry of space-time, and therefore so is $\mathbf{B}^{(3)}$, which as we shall see, emerges directly from the Dirac equation of e in A_μ.

The isomorphism between SU(2) and O(3) in space is extended to one between SL(2,C) and the Lorentz group in space-time. In space,

$$U = e^{i\boldsymbol{\sigma}\cdot\boldsymbol{\theta}/2} \leftrightarrow R = e^{i\mathbf{J}\cdot\boldsymbol{\theta}}, \quad (57)$$

and

$$[\hat{J}_x, \hat{J}_y] = i\hat{J}_z, \text{ et cyclicum,} \quad (58)$$

$$\leftrightarrow \left[\frac{\hat{\sigma}_x}{2}, \frac{\hat{\sigma}_y}{2}\right] = i\frac{\hat{\sigma}_z}{2}, \text{ et cyclicum.}$$

showing that the Pauli matrices become *angular momentum operators* in quantum mechanics. This is the origin of terms such as *spin-half angular momentum*. If $\alpha \to \alpha + 2\pi$ then $U \to -U$; $R \to R$, so R in O(3) can be represented in SU(2) by either U or -U. This finding is summarized in topology [16] by the fact that there is a two to one mapping of the elements of SU(2) on to those of O(3). There is a global topological difference [16] between the two groups.

1.2.2 SPINORS OF THE SL(2,C) GROUP

There is an isomorphism between the SL(2,C) and Lorentz groups which is the geometrical basis of the Dirac equation in space-time, an equation which describes the Lorentz boost transformation in terms of four component spinors. The Lorentz group contains three boost generators (Vol. 1) $\hat{K}_X$, $\hat{K}_Y$ and $\hat{K}_Z$, which can be expressed as 4 x 4 matrices, as well as three rotation generators $\hat{J}_X$, $\hat{J}_Y$ and $\hat{J}_Z$, which are also 4 x 4 matrices. In Vol. 1, the magnetic and electric components of free space electromagnetism were expressed directly in terms of these six generators, both in a Cartesian basis, (X, Y, Z) and a circular basis (1), (2), (3). The O(3) group on the other hand contains only rotation generators which are 3 x 3 matrices in space, rather than 4 x 4 matrices in space-time. The field $B^{(3)}$ in the Lorentz group becomes directly proportional (Vol. 1) to the $\hat{J}_Z$ rotation generator, and therefore it must also have its counterpart in SL(2,C).

The Lorentz group is characterized by a Lie algebra between the boost and rotation generators [16]:

$$[\hat{J}_X, \hat{J}_Y] = i\hat{J}_Z, \qquad \text{et cyclicum,}$$

$$[\hat{K}_X, \hat{K}_Y] = -i\hat{J}_Z, \qquad "\quad",$$

$$[\hat{J}_X, \hat{K}_Y] = i\hat{K}_Z, \qquad "\quad", \tag{59}$$

$$[\hat{J}_X, \hat{K}_X] = 0 \qquad \text{for all X, Y, Z.}$$

This algebra can be re-expressed directly, however [16], as the SU(2) commutators:

$$[\hat{A}_X, \hat{A}_Y] = i\hat{A}_Z, \qquad \text{et cyclicum,}$$

$$[\hat{B}_X, \hat{B}_Y] = i\hat{B}_Z, \qquad "\quad", \tag{60}$$

$$[\hat{A}_j, \hat{B}_j] = 0, \qquad (i, j = X, Y, Z).$$

where

$$\hat{A} := \frac{1}{2}(\hat{J} + i\hat{K}), \qquad \hat{B} := \frac{1}{2}(\hat{J} - i\hat{K}). \tag{61}$$

Since $\hat{A}$ and $\hat{B}$ both obey SU(2) type commutators, the Lorentz group is a direct product group SU(2) ⊗ SU(2) in which the

Geometrical Basis [16] of the Dirac Equation

infinitesimal operators $\hat{A}$ and $\hat{B}$ become complex. In Vol. 1 we saw that the magnetic part ($\hat{B}$) of free space electromagnetism is proportional to $\hat{J}$ of the Lorentz group, and the electric part ($\hat{E}$) to $\hat{K}$. This inference shows that $\hat{A}$ and $\hat{B}$ could equally well be described in terms of complex SU(2) combinations such as,

$$\hat{C} := \frac{1}{2}(\hat{B} + i\hat{E}), \qquad (62a)$$

$$\hat{D} := \frac{1}{2}(\hat{B} - i\hat{E}). \qquad (62b)$$

Equations (62) are the basis for the description of the Maxwell equations with four spinors and Dirac matrices [17]. The Lorentz group is therefore also the SU(2) ⊗ SU(2) group of vacuum electromagnetism, as well as the group of boost and rotation generators. This means that $\hat{B}^{(3)}$ is a real observable, a generator of SU(2) ⊗ SU(2) directly proportional to $\hat{J}_z$. In other words $\hat{B}^{(3)}$ is a space-time field which exists in vacuo, propagates with the transverse fields (Vol. 1) $\hat{B}^{(1)}$ and $\hat{B}^{(2)}$, and is observable through its effect on the trajectory of a single electron, an effect which is characterized by a square root ($I_0^{\frac{1}{2}}$) dependence on the power density of electromagnetic radiation. This characteristic $I_0^{\frac{1}{2}}$ dependence emerges directly from the classical Hamilton-Jacobi equation of e in A_μ (Chap. 12 of Vol. 1) and from the quantum mechanical Dirac equation of e in A_μ (this chapter).

The Lorentz transformation is characterized in general by two different types of two component spinor, which transform as [16]

$$\xi \to M\xi, \qquad \eta \to N\eta, \qquad (63a)$$

where

$$M = e^{i\sigma \cdot \theta/2} e^{\sigma \cdot \phi/2}, \qquad (63b)$$

and

$$N = e^{i\boldsymbol{\sigma}\cdot\boldsymbol{\theta}/2} e^{-\boldsymbol{\sigma}\cdot\boldsymbol{\phi}/2} = \zeta M^* \zeta^{-1}, \qquad (63c)$$

have unit determinants but are not unitary [16]. Under the parity operator

$$\xi \xrightarrow{\hat{P}} \eta. \qquad (64)$$

Here $\boldsymbol{\theta}$ denotes the parameter of a Lorentz rotation [16] and $\boldsymbol{\phi}$ that of a Lorentz boost

$$\boldsymbol{\mathcal{J}} \xrightarrow{\hat{P}} \boldsymbol{\mathcal{J}}, \quad \boldsymbol{\hat{\mathcal{K}}} \xrightarrow{\hat{P}} -\boldsymbol{\hat{\mathcal{K}}}. \qquad (65)$$

Considerations of *parity* therefore lead to the introduction of *the Dirac four-spinor*,

$$\psi := \begin{pmatrix} \xi \\ \eta \end{pmatrix} \xrightarrow{\hat{P}} \begin{pmatrix} 0 & 1 \\ 1 & 0 \end{pmatrix} \begin{pmatrix} \xi \\ \eta \end{pmatrix} = \begin{pmatrix} \eta \\ \xi \end{pmatrix}. \qquad (66)$$

The 2 x 2 matrix representation in this equation is shorthand for a 4 x 4 matrix, because ψ is a four component column vector. Under Lorentz transformation [16],

$$\begin{pmatrix} \xi \\ \eta \end{pmatrix} \rightarrow \begin{pmatrix} e^{\boldsymbol{\sigma}\cdot(\boldsymbol{\theta}-i\boldsymbol{\phi})/2} & 0 \\ 0 & e^{\boldsymbol{\sigma}\cdot(\boldsymbol{\theta}+i\boldsymbol{\phi})/2} \end{pmatrix} \begin{pmatrix} \xi \\ \eta \end{pmatrix} = \begin{pmatrix} D(\Lambda) & 0 \\ 0 & \overline{D}(\Lambda) \end{pmatrix} \begin{pmatrix} \xi \\ \eta \end{pmatrix}, \qquad (67)$$

where

$$\overline{D}(\Lambda) = \zeta D^*(\Lambda) \zeta^{-1}, \qquad (68)$$

(cf. Eq. (63c)). The matrices D and $\overline{D}$ are functions of the Lorentz transformation matrix, Λ, defined in Minkowski notation by

$$x'_\mu = \Lambda_{\mu\nu} x_\nu. \qquad (69)$$

The Dirac four-spinor ψ is an irreducible representation of the product group SU(2) ⊗ SU(2) extended by $\hat{P}$, and from Eqs. (62a) and (62b) can be constructed from complex combinations of magnetic and electric field generators which behave under $\hat{P}$ as

Geometrical Basis [16] of the Dirac Equation

$$\hat{B} \xrightarrow{\hat{P}} \hat{B}, \quad \hat{E} \xrightarrow{\hat{P}} -\hat{E}. \tag{70}$$

Consideration of $\hat{P}$ is therefore of key importance to the Dirac equation, and to the consequent prediction and observation of anti-particles. The helicity as well as the charge of the anti-particle is opposite to that of the original particle.

A pure Lorentz boost is described for finite ϕ by $\theta = 0$, and for $\xi := \phi_R$, $\eta := \phi_L$ [16],

$$\phi_R \rightarrow e^{\sigma \cdot \phi/2} \phi_R = \left(\cosh \frac{\phi}{2} + \sigma \cdot n \sinh \frac{\phi}{2}\right) \phi_R, \tag{71}$$

where n is a unit vector in the direction of the Lorentz boost. *This is essentially one of the Dirac equations of motion for a free particle*; the other being generated by application of $\hat{P}$ to both sides of (71).

1.3 THE FREE PARTICLE DIRAC EQUATION

Considering a Lorentz boost transformation for a particle originally with zero linear momentum in a given frame of reference ($p = 0$, $\phi_R(0)$) to a state where the particle moves with momentum $p(\phi_R(p))$, Eq. (71) becomes

$$\phi_R(p) = \left(\cosh \frac{\phi}{2} + \sigma \cdot n \sinh \frac{\phi}{2}\right) \phi_R(0). \tag{72}$$

Applying $\hat{P}$,

$$\phi_L(p) = \left(\cosh \frac{\phi}{2} - \sigma \cdot n \sinh \frac{\phi}{2}\right) \phi_L(0). \tag{73}$$

The parameter ϕ of the Lorentz boost is given by

$$\cosh \frac{\phi}{2} = \left(\frac{\gamma + 1}{2}\right)^{\frac{1}{2}}, \quad \sinh \frac{\phi}{2} = \left(\frac{\gamma - 1}{2}\right)^{\frac{1}{2}} \tag{74}$$

where, in S.I. units, without suppressing c and h,

$$\gamma = \frac{En}{m_0 c^2}. \tag{75}$$

The two Dirac equations (72) and (73), which interconvert by

20   Chapter 1. $B^{(3)}$ and the Dirac Equation

$\hat{p}$, are thus,

$$\phi_R(p) = \frac{En + m_0 c^2 + c\boldsymbol{\sigma} \cdot \boldsymbol{p}}{\left(2m_0 c^2 (En + m_0 c^2)\right)^{\frac{1}{2}}} \phi_R(0), \tag{76a}$$

$$\phi_L(p) = \frac{En + m_0 c^2 - c\boldsymbol{\sigma} \cdot \boldsymbol{p}}{\left(2m_0 c^2 (En + m_0 c^2)\right)^{\frac{1}{2}}} \phi_L(0), \tag{76b}$$

where $\boldsymbol{p}$ is the relativistic momentum vector in the direction of the Lorentz boost, which, as in Vol. 1, we take as Z of the rest frame K. Note that Eqs. (76a) and (76b) are in S.I. units, whereas most texts in contemporary field theory, including that by Ryder [16], use units in which c and h are suppressed, i.e., are normalized, or set to unity. The energy in these equations is the total relativistic energy, given by

$$En = T + m_0 c^2 = \gamma m_0 c^2, \tag{77}$$

where T is the relativistic kinetic energy [27]. To convert from normalized units, such as those of Ryder, to S.I.,

$$m \rightarrow m_0 c^2, \quad \boldsymbol{p} \rightarrow c\boldsymbol{p}, \quad p_0 := m_0 c, \quad \text{etc.} \tag{78}$$

If the particle is in a frame in which there is no relativistic linear momentum in the Z direction, its helicity vanishes, and therefore there can be no distinction between left and right hand spinors,

$$\phi_R(0) = \phi_L(0). \tag{79}$$

The Dirac equations can therefore be written in terms of the four-spinor ψ as

$$\begin{bmatrix} -m_0 c^2 & c(p_0 + \boldsymbol{\sigma} \cdot \boldsymbol{p}) \\ c(p_0 - \boldsymbol{\sigma} \cdot \boldsymbol{p}) & -m_0 c^2 \end{bmatrix} \begin{pmatrix} \phi_R(p) \\ \phi_L(p) \end{pmatrix} = \begin{pmatrix} 0 \\ 0 \end{pmatrix}, \tag{80}$$

where $m_0 c^2 = p_0 c$, $p_0 = m_0 c$. In terms of the dimensionless 4×4 Dirac matrices

The Free Particle Dirac Equation

$$\gamma_0 := \begin{pmatrix} 0 & 1 \\ 1 & 0 \end{pmatrix}, \quad \gamma_i := \begin{pmatrix} 0 & -\sigma_i \\ \sigma_i & 0 \end{pmatrix}, \tag{81}$$

Eq. (80) is

$$(\gamma_0 c p_0 + \gamma_i c p_i - m_0 c^2)\psi(\mathbf{p}) = 0. \tag{82}$$

Note that the Dirac matrices are written conventionally as two by two square matrices, but each element is a two by two matrix, the off diagonals being Pauli matrices.

In Minkowski notation,

$$\gamma_\mu p_\mu = \gamma_i p_i - \gamma_0 p_0, \tag{83}$$

and the Dirac equation of motion becomes

$$\gamma_\mu p_\mu \psi(\mathbf{p}) = -m_0 c \psi(\mathbf{p}), \tag{84}$$

where p_μ, the momentum-energy four-vector is, in Minkowski notation,

$$p_\mu = \left(\mathbf{p}, i\frac{En}{c}\right) := (p_i, ip_0), \tag{85}$$

and the four-vector of Dirac matrices is, in the same notation,

$$\gamma_\mu := (\gamma_i, i\gamma_0). \tag{86}$$

The minus sign in Eq. (84) is the result of our use of Minkowski notation, which is used for the sake of clarity, and to aid comprehension for non-specialists who wish to understand the origins of $\mathbf{B}^{(3)}$ in the Dirac equation and who may be unfamiliar with the contravariant covariant notation of field theory. The physical meaning of the equation is of course independent of space-time notation and units.

The Dirac equation (84) is more accurately described as the Dirac equations, and this point is emphasized when the rest mass of the particle is zero. In this case we recover the decoupled Weyl equations of motion,

$$(p_0 + \boldsymbol{\sigma}\cdot\mathbf{p})\phi_L(\mathbf{p}) = 0, \quad (p_0 - \boldsymbol{\sigma}\cdot\mathbf{p})\phi_R(\mathbf{p}) = 0, \tag{87}$$

traditionally used for the neutrino. The helicity of the neutrino is defined then by

$$\lambda = \frac{\pm \boldsymbol{\sigma} \cdot \boldsymbol{p}}{|\boldsymbol{\sigma} \cdot \boldsymbol{p}|}. \tag{88}$$

For our present purposes, note that the Dirac equation quantizes into

$$i\gamma_\mu \frac{\partial}{\partial x_\mu} \psi = \frac{m_0 c}{\hbar} \psi, \tag{89}$$

using Eqs. (3), the fundamental axioms of quantum mechanics. The Dirac equation (89) of a free particle is an eigen-equation of quantum mechanics, with wave-function ψ. It leads to a positive definite probability density as demonstrated below, and can therefore be interpreted as a free particle equation in relativistic quantum mechanics. In demonstrating the existence of $B^{(3)}$ from the Dirac equation, the particle is the electron, e, interacting with A_μ.

Applying the operator $i\gamma_\mu \frac{\partial}{\partial x_\mu}$ to both sides of Eq. (89),

$$\gamma_\mu \gamma_\nu \frac{\partial}{\partial x_\mu} \frac{\partial}{\partial x_\nu} \psi = -\left(\frac{m_0 c}{\hbar}\right)^2 \psi, \tag{90}$$

and using the definition of the d'Alembertian,

$$\hat{\Box} := \frac{\partial}{\partial x_\nu} \frac{\partial}{\partial x_\mu}, \tag{91}$$

and the notation

$$[\gamma_\mu, \gamma_\nu] := \gamma_\mu \gamma_\nu + \gamma_\nu \gamma_\mu, \tag{92}$$

Eq. (90) becomes

$$\frac{1}{2}[\gamma_\mu, \gamma_\nu]\hat{\Box}\psi = -\left(\frac{m_0 c}{\hbar}\right)^2 \psi. \tag{93}$$

However, the energy momentum mass relation, Eq. (1), must apply to this analysis, because the overall aim is to find a

The Free Particle Dirac Equation

physically acceptable equivalent of Eq. (1) in quantum mechanics. Each component of the four-spinor ψ must therefore satisfy Eq. (1). This is possible if and only if the Dirac matrices satisfy the relation

$$[\gamma_\mu, \gamma_\nu] = 2\delta_{\mu\nu}, \tag{94}$$

where $\delta_{\mu\nu}$ is the unit matrix for $\mu = \nu$ and is zero otherwise. This allows the Dirac equation to be written in the same form as the Proca equation,

$$\hat{\Box}\psi = -\left(\frac{m_0 c}{\hbar}\right)^2 \psi, \tag{95}$$

which is Eq. (17).

1.3.1 PROBABILITY CURRENT AND DENSITY FROM THE DIRAC EQUATION

The Dirac equation gives a probability density which is positive definite, and for this reason is an acceptable particle equation in quantum mechanics. The probability four-current j_μ, analogous to Eq. (9) is constructed by taking the Hermitian conjugate of Eq. (89), noting that

$$\gamma_0^+ = \gamma_0, \qquad \gamma_i^+ = \gamma_i. \tag{96}$$

This procedure gives the Dirac equation,

$$\psi^+\left(i\gamma_\mu^+ \frac{\overleftarrow{\partial}}{\partial x_\mu^+} - \frac{m_0 c}{\hbar}\right) = 0, \tag{97}$$

in S.I. units. Here ψ^+ is a row vector, and the operator $\overleftarrow{\partial}/\partial x_\mu^+$ operates to the left on this row vector [16]. Multiplying by γ_0 and using

$$\gamma_i \gamma_0 = -\gamma_0 \gamma_i, \tag{98}$$

gives the Dirac equation

$$\overline{\psi}\left(i\gamma_\mu \overset{\leftarrow}{\frac{\partial}{\partial x_\mu}} + \frac{m_0 c}{\hbar}\right) = 0, \qquad (99)$$

where $\overline{\psi}$ is the *Dirac adjoint spinor*,

$$\overline{\psi} = \psi^+ \gamma_0. \qquad (100)$$

The probability current four-vector from the Dirac equation is defined as the unitless expectation value,

$$j_\mu^{(D)} = \overline{\psi}\gamma_\mu \psi. \qquad (101)$$

The four-vector $j_\mu^{(D)}$ is conserved because

$$\frac{\partial j_\mu^{(D)}}{\partial x_\mu} = \frac{\partial}{\partial x_\mu}\left(\overline{\psi}\gamma_\mu \psi\right), \qquad (102)$$

but

$$i\gamma_\mu \frac{\partial \psi}{\partial x_\mu} = \frac{m_0 c}{\hbar}\psi, \text{ and similarly for } \overline{\psi}. \qquad (103)$$

Finally therefore

$$\frac{\partial j_\mu^{(D)}}{\partial x_\mu} = -i\frac{m_0 c}{\hbar}\psi\overline{\psi} + i\overline{\psi}\psi\frac{m_0 c}{\hbar} = 0. \qquad (104)$$

Moreover, the probability density, the time-like component of j_μ, is given by

$$j_0^{(D)} = \overline{\psi}\gamma_0 \psi = \psi^+\psi, \qquad (105)$$

i.e., is simply the product of the Dirac spinor with its own Hermitian transpose. This product is positive definite in

The Free Particle Dirac Equation

the theory of complex numbers [28]. If the components of the spinor are denoted ψ_1, ψ_2, ψ_3 and ψ_4, following Ryder [16], then

$$\psi^+\psi = |\psi_1|^2 + |\psi_2|^2 + |\psi_3|^2 + |\psi_4|^2, \tag{106}$$

which is the sum of squares of moduli, i.e., must be positive definite and rigorously non-negative.

1.3.2 ENERGY EIGENVALUES OF THE DIRAC EQUATION

Considering a particle in a frame of reference in which it is at rest, with zero Z axis linear momentum, therefore, the Dirac equation becomes, setting $\boldsymbol{p} = 0$,

$$\gamma_0 p_0 \psi = m_0 c \psi. \tag{107}$$

Multiplying both sides by γ_0,

$$p_0 \psi = \gamma_0 m_0 c \psi, \tag{108}$$

which is a quantum mechanical wave equation. However, the wave function ψ is a four component spinor, and so we have four wave equations, and four eigenvalues of p_0. Since rest energy is cp_0, there are four eigenvalues of the rest energy operator cp_0. By definition of the 4 x 4 matrix γ_0, the eigenvalues of rest energy from Eq. (108) are $+m_0c^2$, $+m_0c^2$, $-m_0c^2$, and $-m_0c^2$. There are two positive and two negative eigenvalues of rest energy from the Dirac equation of motion. The two positive eigenvalues correspond to the degenerate energy eigenstates of the spin 1/2 particle. However, there are also two corresponding *negative energy eigenstates*.

Dirac circumvented this difficulty through his postulate of anti-particles and through the postulate of the Dirac sea, as described in numerous texts. For our present purposes we note simply that anti-particles have been well verified experimentally, making the Dirac equation one of the most powerful and generally applicable of all equations of motion. Its prediction of $\boldsymbol{B}^{(3)}$ is therefore based firmly in well accepted fundamental relativistic quantum theory.

1.3.3 STANDARD REPRESENTATION OF THE DIRAC EQUATION

Contemporary field theory employs the standard representation of four component Dirac spinors and Dirac matrices, a representation which is briefly explained to complete our survey of the Dirac equation of a free particle. In the next section, the standard representation will be used to derive $B^{(3)}$ directly from the Dirac equation of a particle, the electron, in the classical electromagnetic field A_μ. This indicates that $B^{(3)}$ is also a new fundamental field in quantum electrodynamics where the field is quantized with path integral formalism [16].

In the rest frame, with $p=0$ and $En = m_0 c^2$, the Dirac equation becomes

$$\hat{p}^{(0)} \psi(p) = \gamma_0 m_0 c \psi(p), \qquad (109)$$

which is a first order partial differential equation in ψ. The solutions can be written in the forms [16]

$$\psi = u(0) \exp\left(-i \frac{m_0 c^2}{\hbar} t\right), \qquad (110a)$$

$$\psi = v(0) \exp\left(i \frac{m_0 c^2}{\hbar} t\right), \qquad (110b)$$

which are respectively the positive and negative energy solutions. These are plane wave solutions with positive and negative energy spinors:

$$u^{(1)}(0) = \begin{bmatrix} 1 \\ 0 \\ 0 \\ 0 \end{bmatrix}, \quad u^{(2)}(0) = \begin{bmatrix} 0 \\ 1 \\ 0 \\ 0 \end{bmatrix}, \quad v^{(1)}(0) = \begin{bmatrix} 0 \\ 0 \\ 1 \\ 0 \end{bmatrix},$$

$$v^{(2)}(0) = \begin{bmatrix} 0 \\ 0 \\ 0 \\ 1 \end{bmatrix}, \qquad (111)$$

in the *standard representation where γ_0 is diagonal*

The Free Particle Dirac Equation

$$\gamma_0^{(SR)} = \begin{pmatrix} 1 & 0 & 0 & 0 \\ 0 & 1 & 0 & 0 \\ 0 & 0 & -1 & 0 \\ 0 & 0 & 0 & -1 \end{pmatrix}. \tag{112}$$

In condensed form

$$\gamma_0^{(SR)} = \begin{pmatrix} 1 & 0 \\ 0 & -1 \end{pmatrix} = S\gamma_0 S^{-1}, \tag{113}$$

where

$$S = \frac{1}{\sqrt{2}} \begin{pmatrix} 1 & 1 \\ 1 & -1 \end{pmatrix}. \tag{114}$$

Therefore,

$$\gamma_0^{(SR)} S = S\gamma_0, \tag{115}$$

and the four-spinor is given by

$$\psi = S \begin{pmatrix} \phi_R \\ \phi_L \end{pmatrix} = \frac{1}{\sqrt{2}} \begin{pmatrix} \phi_R + \phi_L \\ \phi_R - \phi_L \end{pmatrix}. \tag{116}$$

The equivalent of the Lorentz boost transformation matrix in standard representation is

$$M^{(SR)} = SMS^{-1} = \begin{pmatrix} \cosh\frac{\phi}{2} & \boldsymbol{\sigma}\cdot\boldsymbol{n}\sinh\frac{\phi}{2} \\ \boldsymbol{\sigma}\cdot\boldsymbol{n}\sinh\frac{\phi}{2} & \cosh\frac{\phi}{2} \end{pmatrix}. \tag{117}$$

The plane wave spinors are therefore

$$\psi^{(\alpha)}(Z) = u^{(\alpha)}(\boldsymbol{p})\exp\left(-i\frac{p}{\hbar}Z\right), \tag{118a}$$

$$\psi^{(\alpha)}(Z) = v^{(\alpha)}(\boldsymbol{p})\exp\left(i\frac{p}{\hbar}Z\right), \tag{118b}$$

which are respectively the positive and negative energy solutions, with $\alpha = 1$ and 2 in each case.

Specifically, in S.I. units:

$$u^{(1)}(\boldsymbol{p}) = M^{(SR)} u^{(1)}(\boldsymbol{0}) = \left(\frac{\zeta}{2m_0 c^2}\right)^{\frac{1}{2}} \begin{bmatrix} 1 \\ 0 \\ \frac{p_z c}{\zeta} \\ \frac{(p_x + ip_y)c}{\zeta} \end{bmatrix},$$

$$u^{(2)}(\boldsymbol{p}) = M^{(SR)} u^{(2)}(\boldsymbol{0}) = \left(\frac{\zeta}{2m_0 c^2}\right)^{\frac{1}{2}} \begin{bmatrix} 0 \\ 1 \\ \frac{(p_x - ip_y)c}{\zeta} \\ \frac{-p_z c}{\zeta} \end{bmatrix},$$

(119)

$$v^{(1)}(\boldsymbol{p}) = M^{(SR)} v^{(1)}(\boldsymbol{0}) = \left(\frac{\zeta}{2m_0 c^2}\right)^{\frac{1}{2}} \begin{bmatrix} \frac{p_z c}{\zeta} \\ \frac{(p_x + ip_y)c}{\zeta} \\ 1 \\ 0 \end{bmatrix},$$

$$v^{(2)}(\boldsymbol{p}) = M^{(SR)} v^{(2)}(\boldsymbol{0}) = \left(\frac{\zeta}{2m_0 c^2}\right)^{\frac{1}{2}} \begin{bmatrix} \frac{(p_x - ip_y)c}{\zeta} \\ \frac{-p_z c}{\zeta} \\ 0 \\ 1 \end{bmatrix},$$

where $\zeta := En + m_0 c^2$.

1.4 THE DIRAC EQUATION OF e IN A_μ: PROOF OF $B^{(3)}$ FROM FIRST PRINCIPLES

The Dirac equation of e in A_μ is obtained by replacing p_μ in Eq. (84) by the sum $p_\mu + eA_\mu$, where $A_\mu := \left(\mathbf{A}, \frac{i\phi}{c}\right)$. This is the well known *minimal prescription*, whose origins can be found in gauge invariance of the second kind (e.g. Refs. [16] and [17] and Vol. 1). In this view, electromagne-

Dirac Equation, Proof of $B^{(3)}$ From First Principles

tism is the product eA_μ that keeps the action invariant under gauge transformation of the second kind in the presence of the field. The gauge potential therefore couples to the four-current J_μ with strength e, which is the *charge* of the scalar field ϕ. The derivative

$$D_\mu \phi := \left(\frac{\partial}{\partial x_\mu} + \frac{ieA_\mu}{\hbar} \right) \phi \tag{120}$$

transforms covariantly under gauge transformation, as does the scalar field itself.

The Dirac equation of e in A_μ is therefore

$$\gamma_\mu (p_\mu + eA_\mu) \psi(p) = -m_0 c \psi(p), \tag{121}$$

or in vector notation

$$(\gamma_0(En + e\phi) - c\boldsymbol{\gamma} \cdot (\boldsymbol{p} + e\boldsymbol{A}))\psi = m_0 c^2 \psi. \tag{122}$$

In the standard representation this splits into

$$(En + e\phi)u - c\boldsymbol{\sigma} \cdot (\boldsymbol{p} + e\boldsymbol{A})v = m_0 c^2 u, \tag{123a}$$

$$-(En + e\phi)v + c\boldsymbol{\sigma} \cdot (\boldsymbol{p} + e\boldsymbol{A})u = m_0 c^2 v, \tag{123b}$$

from the second of which

$$v = \left(\frac{c\boldsymbol{\sigma} \cdot (\boldsymbol{p} + e\boldsymbol{A})}{En + m_0 c^2 + e\phi} \right) u. \tag{124}$$

In the rest frame approximation

$$En \sim m_0 c^2, \tag{125}$$

(i.e., when there is no net linear Z axis electron momentum),

$$v \sim \left(\frac{c\boldsymbol{\sigma} \cdot (\boldsymbol{p} + e\boldsymbol{A})}{2m_0 c^2 + e\phi} \right) u. \tag{126}$$

So Eq. (124) becomes, with $W := E_n - m_0 c^2$, $\mathbf{\Pi} := \mathbf{p} + e\mathbf{A}$,

$$\hat{W}u = \left(\frac{(\mathbf{\sigma} \cdot \mathbf{\Pi})(\mathbf{\sigma} \cdot \mathbf{\Pi})}{2m_0 + e\phi/c^2} - e\phi \right) u, \qquad (127)$$

which becomes

$$\hat{W}u = \left(\frac{(\mathbf{\sigma} \cdot \mathbf{\Pi})(\mathbf{\sigma} \cdot \mathbf{\Pi})}{2m_0} - e\phi \right) u, \qquad (128)$$

in the limit

$$e\phi \ll 2m_0 c^2. \qquad (129)$$

1.4.1 EMERGENCE OF $B^{(3)}$ FROM EQUATION (128)

Equation (128) is a wave equation of quantum mechanics which gives the Hamiltonian eigenvalue

$$\hat{W}u = Hu, \qquad (130a)$$

$$H = \frac{(\mathbf{\sigma} \cdot \mathbf{\Pi})^2}{2m_0} - e\phi, \qquad (130b)$$

with, from spinor algebra [16],

$$(\mathbf{\sigma} \cdot \mathbf{\Pi})^2 = (\mathbf{p} + e\mathbf{A})^2 + i\mathbf{\sigma} \cdot (\mathbf{p} + e\mathbf{A}) \times (\mathbf{p} + e\mathbf{A}). \qquad (131)$$

Since H is a Hamiltonian, it is a *constant* of the motion of the electron e in the classical electromagnetic field, represented by A_μ. In Chap. 12 of Vol. 1 the Hamilton-Jacobi equation was used to show that in a frame of reference in which the net electronic angular momentum is zero, the classical electron trajectory in A_μ is a circle, generated by the field $B^{(3)}$. The classical Hamiltonian, furthermore, is

$$H_{class} = \frac{1}{2m_0} (\mathbf{p} + e\mathbf{A})^2 - e\phi. \qquad (132)$$

Dirac Equation, Proof of $B^{(3)}$ From First Principles

The Hamiltonian

$$H_{spin} = \frac{i\boldsymbol{\sigma}}{2m_0} \cdot (\boldsymbol{p} + e\boldsymbol{A}) \times (\boldsymbol{p} + e\boldsymbol{A}), \quad (133)$$

therefore occurs in addition to H_{class} and is inherently quantum mechanical in nature. By considering $\boldsymbol{A}$ to be a vector, and $\hat{\boldsymbol{p}}$ to be a vector *operator*, $\hat{\boldsymbol{p}} = -i\hbar\hat{\nabla}$, of quantum mechanics, the following commutators are obtained:

$$[\hat{p}_i, A_j] = \hat{p}_i A_j - A_j \hat{p}_i = -\frac{i\hbar \partial A_j}{\partial x_i}, \quad (134a)$$

$$[A_i, \hat{p}_j] = A_i \hat{p}_j - \hat{p}_j A_i = \frac{i\hbar \partial A_i}{\partial x_j}, \quad (134b)$$

where the terms $A_j \hat{p}_i$ and $A_i \hat{p}_j$ have been defined to be zero. This means that $\hat{p}$ operates on $\boldsymbol{A}$ but $\boldsymbol{A}$ does not operate on $\hat{p}$. Adding (134a) and (134b) gives [16]

$$-i\hbar B_k = -i\hbar \left(\frac{\partial A_j}{\partial x_i} - \frac{\partial A_i}{\partial x_j} \right) = [\hat{p}_i, A_j] + [A_i, \hat{p}_j], \quad (135)$$

where B_k is the *magnetic flux density*,

$$B_k = \frac{i}{\hbar} ([\hat{p}_i, A_j] + [A_i, \hat{p}_j]). \quad (136)$$

The various terms in H_{spin} can be developed in more detail with this result.

In general, the vector potential $\boldsymbol{A}$ of an electromagnetic field is a complex quantity (Vol. 1) and in the circular basis (1), (2), (3) of that volume can be represented by $\boldsymbol{A}^{(1)}$ and its complex conjugate $\boldsymbol{A}^{(2)}$. These are transverse plane waves, through which the usual transverse magnetic wave fields can be defined,

$$\boldsymbol{B}^{(1)} = \nabla \times \boldsymbol{A}^{(1)}, \quad \boldsymbol{B}^{(2)} = \nabla \times \boldsymbol{A}^{(2)}. \quad (137)$$

Conjugate products such as $\boldsymbol{A}^{(1)} \times \boldsymbol{A}^{(2)}$ in Eq. (133) are therefore non-zero, and contribute to H_{spin}. These are considered in the next section. When A is pure real, however, it follows [16] that Eq. (133) can be expressed as

$$H_{spin} = \frac{e\hbar}{2m_0} \boldsymbol{\sigma} \cdot \boldsymbol{B}, \tag{138}$$

where B is a magnetic flux density defined by the operator sum [16],

$$\boldsymbol{B} := \nabla \times \boldsymbol{A} = \frac{i}{\hbar}(\hat{\boldsymbol{p}} \times \boldsymbol{A} + \boldsymbol{A} \times \hat{\boldsymbol{p}}). \tag{139}$$

In this equation, $\hat{p}$ is a vector operator of quantum mechanics, not a classical momentum vector. It defines *the intrinsic (irremovable) transverse linear momentum operator of the electron, corresponding to the intrinsic spin angular momentum operator, whose non-zero eigenvalue is the universal Dirac constant* $\hbar$ *of quantum mechanics*. The Hamiltonian H_{spin} is therefore to order one half in the power density of the classical electromagnetic field (watts per square metre). It is proportional to $B^{(3)}$ at order one, and to the electronic spin angular momentum, $\hbar\boldsymbol{\sigma}$, at order one. The concept of intrinsic electronic spin does not exist in classical physics, because there $\hbar \to 0$, and therefore the transverse components of $\hat{p}$ appearing in Eq. (139) also disappear in classical physics. On the other hand, the vector A is a classical vector potential, and is non-zero in classical physics.

Equation (138) shows that the magnetic flux density B appearing in the spin part of the Hamiltonian, H_{spin}, is *independent of time*. The only time-independent (i.e., phase free) magnetic component of vacuum electrodynamics is $B^{(3)}$ (Vol. 1) and so

$$H_{spin} = \frac{e\hbar}{2m_0} \boldsymbol{\sigma} \cdot \boldsymbol{B}^{(3)} = \pm \frac{e\hbar B^{(0)}}{2m_0} \tag{140}$$

for the interaction of e with A_μ. For one sense of circular polarization (e.g. right (R))

$$H_{spin}^{(R)} = \pm \frac{e\hbar B^{(0)}}{2m_0}, \tag{141}$$

and for the other (left (L))

Dirac Equation, Proof of $B^{(3)}$ From First Principles

$$H_{spin}^{(L)} = \mp \frac{e\hbar B^{(0)}}{2m_0}, \qquad (142)$$

because the sign of $B^{(3)}$ changes when the sense of circular polarization is reversed.

Equation (140) is a fundamental, first principles, demonstration of the existence of $B^{(3)}$ as an observable of free space electromagnetism, because H_{spin} is a direct result of the Dirac equation of e in A_μ. It shows that circularly polarized electromagnetic radiation generates a magnetic field in vacuo, a field which is the fundamental entity of magneto-optics.

In deriving this result, use has been made of the operator-vector definitions [16],

$$(\hat{p} \times A)_k := -i\hbar \frac{\partial A_j}{\partial x_i}, \qquad \left(A \times \hat{p}\right)_k := i\hbar \frac{\partial A_i}{\partial x_j}, \qquad (143)$$

which link Eqs. (136) and (139).

The importance of the result (140) for magnetic effects of electromagnetic radiation cannot be overemphasized, because $B^{(3)}$ is the fundamental magnetizing field. In Eq. (140) it is viewed as a classical field, because the Dirac equation describes the trajectory of e in the classical A_μ. The spin Hamiltonian H_{spin} is built up from the dot product of $B^{(3)}$ with $e\hbar\sigma/(2m_0)$ which disappears in the classical limit $\hbar \to 0$. Therefore H_{spin} has no classical counterpart for this reason, despite the fact that $B^{(3)}$ is a non-zero, classical field. Intrinsic electron spin has been well known and used for over sixty years, but the simple additional inference respresented by Eq. (140) appears not to have been made, despite the fact that *both* the electron spin *and* $B^{(3)}$ are derived *simultaneously* from the same equation of motion of one electron in the classical electromagnetic field.

1.4.2 COMPLEX A_μ: SECOND ORDER PROCESS

Since H_{spin} in Eq. (140) is a Hamiltonian, it is time independent, showing that $B^{(3)}$ is a phase free, time independent, and observable component of vacuum electrodynamics. Equation (4) of Vol. 1, furthermore, relates $B^{(3)}$ to the plane waves $B^{(1)}$ and $B^{(2)}$, which are complex conjugates in the basis

(1), (2), and (3) of Vol. 1. These magnetic plane waves are defined by Eqs. (70) and (71) of that volume. In these equations (Vol. 1)

$$\mathbf{A}^{(1)} = \frac{cB^{(0)}}{\sqrt{2}\omega}(i\mathbf{i} + \mathbf{j})e^{i\phi}, \quad \mathbf{A}^{(2)} = \frac{cB^{(0)}}{\sqrt{2}\omega}(-i\mathbf{i} + \mathbf{j})e^{-i\phi}, \quad (144)$$

and it follows from the minimal prescription,

$$p_\mu \to p_\mu + eA_\mu \tag{145}$$

for the motion of e in A_μ, that the transverse momenta of the electron are in general complex operators, represented in the same basis by the complex conjugate pairs

$$\mathbf{p}^{(1)} = \mathbf{p}^{(2)*}. \tag{146}$$

In so doing, it is understood that measurable quantities (physical observables) are real, as in electrodynamics in general. Thus, for example, the conjugate products $\mathbf{A}^{(1)} \times \mathbf{A}^{(2)}$ and $\mathbf{p}^{(1)} \times \mathbf{p}^{(2)}$ are pure imaginary in the representation (144), but contribute to H_{spin} in Eq. (133) by multiplying the imaginary $i\sigma/(2m_0)$. Dimensionality shows, furthermore, that these conjugate products must be *phase free, magnetic fields* akin to $\mathbf{B}^{(3)}$, i.e., relations such as (Chaps. 3 and 4)

$$\mathbf{B}_1^{(3)} = i\frac{e}{\hbar}(\mathbf{A}^{(1)} \times \mathbf{A}^{(2)}) \tag{147}$$

are expected, where $\mathbf{B}_1^{(3)}$ is a magnetic field. Using Eq. (12) of Vol. 1

$$\mathbf{A}^{(1)} \times \mathbf{A}^{(2)} = i\left(\frac{c}{\omega}\right)^2 B^{(0)} \mathbf{B}^{(3)}, \tag{148}$$

and so

$$\mathbf{B}_1^{(3)} = -\frac{e}{\hbar}\left(\frac{c}{\omega}\right)^2 B^{(0)} \mathbf{B}^{(3)}. \tag{149}$$

Assuming that $\mathbf{B}_1^{(3)} = \mathbf{B}^{(3)}$, then without loss of generality we arrive at

Dirac Equation, Proof of $B^{(3)}$ From First Principles

$$eB^{(0)} = -\hbar \boldsymbol{\kappa} \cdot \boldsymbol{\kappa} = -\hbar \frac{\omega^2}{c^2} \tag{150}$$

in order for Eq. (147) to be valid.

To interpret Eqs. (147) and (149) requires that

$$e\mathbf{A}^{(1)} \rightarrow \left(\hat{\mathbf{p}}^{(1)} = \hbar \hat{\boldsymbol{\kappa}}^{(1)} := -i\hbar \hat{\boldsymbol{\nabla}}^{(1)} \right), \tag{151}$$

i.e., the *electron* momentum operator $\hat{\mathbf{p}}^{(1)}$ is generated from the product of the electronic charge e and the field vector potential A_μ. Anticipating a quantized field interpretation, then the *electron* momentum is generated from the *photon* momentum operator corresponding to the classical $\mathbf{A}^{(1)}/e$. This is an example of an electron property being created from the equivalent photon property. In order therefore to interpret Eq. (150),

$$(\hbar \boldsymbol{\kappa})_{photon} \rightarrow (\hat{p})_{electron}. \tag{152}$$

From Eq. (151) in Eq. (147)

$$\mathbf{B}_1^{(3)} = \boldsymbol{\nabla}^{(1)} \times \mathbf{A}^{(2)}, \tag{153}$$

which is formally identical with the usual classical equation of $\mathbf{B}$ with $\boldsymbol{\nabla} \times \mathbf{A}$.

By considering the possibility of a non-zero $\mathbf{A}^{(1)} \times \mathbf{A}^{(2)}$ in the Dirac equation, the meaning of Eq. (12) of Vol. 1 can be clarified in terms of Eqs. (150) and (152), in which momentum is taken from the classical field and used to create the electron momentum operator $\hat{p}$, which is finally quantized according to the fundamental axiom used in Eq. (151). With this prescription, Eq. (147) becomes formally identical with Eq. (139), but physically, the electron momentum $\hat{p}$ in Eq. (139) is not obtained from photon momentum. The Dirac spin Hamiltonian from Eq. (147) is therefore

$$H_{spin, 1} = \frac{e\hbar}{2m_0} \boldsymbol{\sigma} \cdot \mathbf{B}_1^{(3)}, \tag{154}$$

which can be written as

$$H_{spin, 1} = -\left(\frac{\hbar\kappa}{2m_0}\right)_{electron} \cdot (\hbar\kappa)_{photon} \qquad (155)$$

to emphasize the fact that electron spin has been captured from the electromagnetic field itself, a process which is second order in A, as Eq. (147) indicates on the right hand side.

The existence of $B^{(3)}$ from Eq. (12) of Vol. 1 therefore leads consistently to the second order Dirac spin Hamiltonian (155), in which the photon (i.e., the quantized field) representation has been anticipated.

1.4.3 COMPLEX A_μ: FIRST ORDER PROCESS

The first order process with real A leads to the spin Hamiltonian (140), in which the presence of the classical electromagnetic field is represented solely by $B^{(3)}$, which is in turn proportional to the square root of field intensity (i.e., power density). The process just described in section 1.4.2. is, on the other hand, to order one in power density. The field $B^{(3)}$ is formed to first order from

$$B^{(3)} = \frac{i}{\hbar}\left((\hat{p}^{(1)} \times A^{(2)})_Z + \left(A^{(2)} \times \overleftarrow{\hat{p}}^{(1)}\right)_Z\right)e^{(3)}, \qquad (156)$$

which is formally identical with Eq. (139). However, in Eq. (156), the momentum vector operator $\hat{p}$ is considered to be a complex quantity in the basis (1), (2), (3), and therefore the operator

$$\partial_i = \frac{i\hat{p}_i}{\hbar} \qquad (157)$$

now has real and imaginary components. The physically meaningful part of $B^{(3)}$ from Eq. (156) is its *real* part, and Eq. (156) reduces to

$$B^{(3)} = \frac{i}{\hbar}\left(\hat{p}_X^{(1)} A_Y^{(2)} - \hat{p}_Y^{(1)} A_X^{(2)}\right)e^{(3)} \qquad (158)$$

using the semi-commutators [16]

Dirac Equation, Proof of $B^{(3)}$ From First Principles

$$(\hat{p}^{(1)} \times A^{(2)})_k := -i\hbar \frac{\partial A_j^{(2)}}{\partial x_i^{(1)}} = \left[\hat{p}_i^{(1)}, A_j^{(2)}\right], \tag{159}$$

$$\left(A^{(2)} \times \hat{p}^{(1)}\right)_k := i\hbar \frac{\partial A_i^{(2)}}{\partial x_j^{(1)}} = \left[A_i^{(2)}, \hat{p}_j^{(1)}\right].$$

Equation (158) indicates the observation of $B^{(3)}$ through the intrinsic electronic spin, and the first order spin Hamiltonian (140). The latter is to first order in the magnetic flux density magnitude, $B^{(0)}$, of the classical electromagnetic field. It is a non-zero, time invariant quantity (a Hamiltonian eigenvalue) and therefore independent of the electromagnetic phase. It must therefore originate in the interaction of $B^{(3)}$ with the quantized electron spin. The only way in which intrinsic electron spin can be observed in this context is through the observable field $B^{(3)}$, for example in a Zeeman splitting due to $B^{(3)}$ of a pump laser. Such an experiment indicates the presence of $B^{(3)}$ in vacuo, and at first order the splitting should be proportional to the square root of the pump laser's power density.

1.5 COMPARISON WITH THE CLASSICAL EQUATION OF MOTION OF e IN A_μ

From Eq. (132) the Hamiltonian eigenvalue of e in A_μ in the Dirac equation also contains the term

$$H_{class} = \frac{1}{2m_0}(\hat{p} + eA)\cdot(\hat{p} + eA), \tag{160}$$

where $V = -Yi + Xj$ is the same operator as for the spin term. Unlike the latter, Eq. (160) has a classical equivalent [15]. If A_3 is the vector potential of $B^{(3)}$ (Vol. 1), then [15]

$$B^{(3)} = \nabla \times A_3, \quad A_3 = \frac{1}{2} B^{(3)} \times r := \frac{1}{2} B^{(0)} V, \tag{161}$$

$$(r := X\mathbf{1} + Y\mathbf{j} + Z\mathbf{k}),$$

where $V = -Y\mathbf{1} + X\mathbf{j}$ is a possible representation in Cartesian coordinates. Therefore $\nabla \cdot A_3 = 0$ and the term in A in Eq. (160) becomes [18]

$$H^{(1)} = \frac{e}{m_0} \boldsymbol{A} \cdot \boldsymbol{p} = \frac{e}{2m_0} \boldsymbol{B}^{(3)} \times \boldsymbol{r} \cdot \boldsymbol{p} = \frac{e}{2m_0} \boldsymbol{B}^{(3)} \cdot \boldsymbol{r} \times \boldsymbol{p}. \qquad (162)$$

In order for this to be non-zero, $\boldsymbol{r} \times \boldsymbol{p}$ must be in the same direction (Z) as $\boldsymbol{B}^{(3)}$. This is possible if and only if both $\boldsymbol{r}$ and $\boldsymbol{p}$ have components *transverse* to the direction of $\boldsymbol{B}^{(3)}$. This reasoning identifies $\boldsymbol{p}$ as transverse electron linear momentum, and identifies $\boldsymbol{r} \times \boldsymbol{p}$ as Eq. (399) of Vol. 1,

$$\boldsymbol{J}^{(3)} = J_Z \boldsymbol{k} = \boldsymbol{r} \times \boldsymbol{p} = (Xp_Y - Yp_X)\boldsymbol{k}, \qquad (163)$$

which is the *orbital* angular momentum of the electron in the classical field $\boldsymbol{B}^{(3)}$. Then

$$H^{(1)} = \frac{e}{2m_0} \boldsymbol{J}^{(3)} \cdot \boldsymbol{B}^{(3)} = -\boldsymbol{m}^{(3)} \cdot \boldsymbol{B}^{(3)}, \qquad (164)$$

where $\boldsymbol{m}^{(3)} = -(e/2m_0) \boldsymbol{J}^{(3)}$ is the induced magnetic dipole moment of Eq. (403) of Vol. 1, obtained from the *classical* Hamilton-Jacobi equation of the motion of e in A_μ. This leads us to expect that there is a Hamilton-Jacobi formulation of the Dirac equation, and this is indeed the case [29]. In Chap. 12 of Vol. 1 we have seen that $\boldsymbol{J}^{(3)}$ is given relativistically by Eq. (402) of that volume,

$$\boldsymbol{J}^{(3)} = \frac{e^2 c^2}{\omega^2} \left(\frac{B^{(0)}}{(m_0^2 \omega^2 + e^2 B^{(0)2})^{\frac{1}{2}}} \right) \boldsymbol{B}^{(3)}, \qquad (165)$$

where ω is the electronic cyclotron frequency set up, i.e., induced, by $\boldsymbol{B}^{(3)}$. On the other hand, the Dirac spin Hamiltonian, H_{spin} of Eq. (140), is the result of the *intrinsic* electronic angular momentum, which is not an induced angular momentum. The complete Hamiltonian in $\boldsymbol{B}^{(3)}$ is therefore

$$H = H^{(1)} + H_{spin} = \frac{e}{2m_0} (\boldsymbol{J}^{(3)} + \hbar \boldsymbol{\sigma}) \cdot \boldsymbol{B}^{(3)}, \qquad (166)$$

and the complete magnetic dipole moment is

$$\boldsymbol{m}^{(3)} = -\frac{e}{2m_0} (\boldsymbol{J}^{(3)} + \hbar \boldsymbol{\sigma}). \qquad (167)$$

Finally, the expected magnetization, if N is the number of electrons in a plasma subjected to irradiation by a circularly polarized electromagnetic field is

$$M^{(3)} = N m^{(3)},\qquad(168)$$

consisting of terms to zeroth, first, and second order in $B^{(3)}$. Equation (168) is the result of the Dirac equation in its Hamilton-Jacobi form. The classical result is Eq. (165) without the term in $\hbar\sigma$, because classically, $\hbar \to 0$, giving Eq. (405) of Vol. 1.

Equation (168) is a simple prediction whose experimental investigation reveals in theory the presence of $B^{(3)}$ through its square root power density dependence. It shows clearly that the Dirac constant, $\hbar$, is an angular momentum (the spinor σ being unitless). In one sense, the angular momentum $\hbar$ of the photon has been given up, or has been transmuted, to that of the electron. It is therefore possible to talk of the photon giving up all its angular momentum to the electron in a process of magnetization by light. This angular momentum obviously has no dependence on field amplitude $B^{(0)}$; for the photon it is $\hbar k$, for the electron it is $\hbar\sigma$. The difference is due to the fact that the photon is a boson, the electron a fermion.

The equation (168) is therefore the rigorous description from first principles of the inverse Faraday effect for N electrons in a plasma.

Chapter 2. $\boldsymbol{B}^{(3)}$ and the Higgs Phenomenon

In Vol. 1 the emergence of $\boldsymbol{B}^{(3)}$ was related to photon mass, the upper limits on which are listed in contemporary standard tables on particle mass [28]. Higgs has suggested [29] that there exists a scalar field, the Higgs field, that gives rise to photon mass through spontaneous symmetry breaking. In this chapter we aim to link the Higgs phenomenon with the field $\boldsymbol{B}^{(3)}$, and to show that the latter is a vortex line, or soliton, in non-linear field theory. In superconductors it becomes an Abrikosov line, but also exists in the vacuum, whose non-trivial topology is well established in the contemporary literature [30].

The cyclically symmetry link between the complex conjugate wave fields $\boldsymbol{B}^{(1)}$ and $\boldsymbol{B}^{(2)}$, and the novel spin field, $\boldsymbol{B}^{(3)}$, is given in Eqs. (4) of Vol. 1, and is

$$\boldsymbol{B}^{(1)} \times \boldsymbol{B}^{(2)} = iB^{(0)}\boldsymbol{B}^{(3)*}, \qquad \boldsymbol{B}^{(2)} \times \boldsymbol{B}^{(3)} = iB^{(0)}\boldsymbol{B}^{(1)*},$$

$$\boldsymbol{B}^{(3)} \times \boldsymbol{B}^{(1)} = iB^{(0)}\boldsymbol{B}^{(2)*}.$$

(169)

These equations are consistent with Maxwell's equations in free space when the mass of the photon (anticipating the quantized interpretation of light) is identically zero. The existence of finite photon mass is not therefore a necessary condition for the existence of $\boldsymbol{B}^{(3)}$. It exists in free space when the photon mass is identically zero. However, the fact that the transverse wave fields imply the existence of the longitudinal spin field through a non-Abelian set of equations (169) suggests that the object known as the photon has three degrees of polarization (1), (2), and (3) in free space. Since photon mass (through the Proca equation, for example) suggests three degrees of space-like polarization there appears to be a link between photon mass and the spin field $\boldsymbol{B}^{(3)}$, which is a real, physical field. The Higgs field [29] also makes the photon massive through spontaneous symmetry breaking, but in the GWS theory of fields [31], a theory which unifies successfully the electromagnetic and

Chapter 2. $B^{(3)}$ and the Higgs Phenomenon

weak fields, the photon mass is forced to be zero by modelling. Nevertheless, GWS is based on the Higgs phenomenon, and on the idea put forward by Higgs of spontaneous symmetry breaking. Contemporary gauge theory asserts that the photon mass is identically zero, but as discussed in Vol. 1, it is easily possible to make the equivalent assertion,

$$A_\mu A_\mu = 0, \quad \left(A_\mu := \left(\mathbf{A}, \, i\frac{A^{(0)}}{c}\right)\right), \tag{170}$$

where A_μ is the four potential. Equation (170) satisfies gauge invariance, while leaving open the possibility of non-zero photon mass. Equation (170), however, implies that

$$|\mathbf{A}| := A^{(0)} \tag{171}$$

and excludes the Coulomb gauge [32], because the time-like part of A_μ is not zero, being equal to the magnitude of the space-like vector potential $\mathbf{A}$. This becomes the condition for the existence of finite photon mass if the latter is not to be identically zero, and if gauge invariance is to be retained as a principle of physics. Finally, fundamental symmetry considerations lead to the conclusion that the cyclically symmetric relations,

$$\mathbf{A}^{(1)} \times \mathbf{A}^{(2)} = -A^{(0)}(i\mathbf{A}^{(3)})^*, \quad \mathbf{A}^{(2)} \times (i\mathbf{A}^{(3)}) = -iA^{(0)}\mathbf{A}^{(1)*}, \tag{1/2}$$

$$(i\mathbf{A}^{(3)}) \times \mathbf{A}^{(1)} = -A^{(0)}\mathbf{A}^{(2)*}.$$

between components of $\mathbf{A}$ can be satisfied if and only if the longitudinal part of $\mathbf{A}$ is pure imaginary. The magnitude of the scalar potential $A^{(0)}$, (the timelike part of A_μ), is therefore

$$|A^{(0)}| = c|\mathbf{A}|, \tag{173}$$

i.e., is the modulus of $\mathbf{A}$, which in the circular basis (1), (2), and (3), is

$$|\mathbf{A}| := (\mathbf{A}^{(1)} \cdot \mathbf{A}^{(1)*} + \mathbf{A}^{(2)} \cdot \mathbf{A}^{(2)*} + \mathbf{A}^{(3)} \cdot \mathbf{A}^{(3)*})^{\frac{1}{2}}. \tag{174}$$

This view excludes the Coulomb gauge because in that gauge it is asserted that the components of **A** are transverse and there is no longitudinal component $A^{(3)}$, and no time-like component, $A^{(0)}$ or scalar potential. For these reasons, the Coulomb gauge is inconsistent with the fact that A_μ is a four-vector, and inconsistent with the simultaneous existence of photon mass and gauge invariance. This conclusion is consistent with the fact that the Proca equation is consistent with the Lorentz gauge condition,

$$\frac{\partial A_\mu}{\partial x_\mu} = 0, \qquad (175)$$

but not with the Coulomb gauge [32].

It is well known that the Higgs phenomenon leads to the inference [29] that the photon becomes massive in superconductors, and to the existence of $B^{(3)}$ as a vortex in topology, recognizable experimentally as a quantized flux, an Abrikosov line. The same Higgs Lagrangian also leads to the Proca equation in superconductors [16]. Mathematically, the Proca equation for the massive photon in free space is the same precisely as the Proca equation for the massive photon, and for the Abrikosov lines, in superconductors. It has been argued in Chap. 12 of Vol. 1 and in the opening chapter of Vol. 2 that the motion of e in A_μ is governed entirely by a free space $B^{(3)}$, a $B^{(3)}$ which is linked to the ordinary transverse $B^{(1)} = B^{(2)*}$ from the standard free space Maxwell equations by the non-Abelian (169). Therefore, if photon mass is accepted as a possibility, $B^{(3)}$ is a topological vortex in free space, and is a consequence of the vacuum topology and spontaneous symmetry breaking. The experimental observation of $B^{(3)}$ in the inverse Faraday effect (Chap. 7 of Vol. 1) would then become evidence for the Higgs field in the vacuum. If photon mass is asserted by axiom to be identically zero, then $B^{(3)}$ becomes a topological consequence of the existence of the Maxwellian $B^{(1)}$ and $B^{(2)}$ through the non-Abelian equations (169). As argued in Vol. 1, if photon mass is asserted to be identically zero, the range of electromagnetism becomes infinite, and therefore exceeds the known dimensions of the universe. It is impossible therefore to test this assertion experimentally. In contrast, the arguments for finite photon mass or classical equivalent have been developed from the time of Cavendish [33], and are ably reviewed in the recent literature [34]. Numerous experiments have been carried out to derive upper limits on finite photon mass, and these limits are available in standard tables. It

Chapter 2. $B^{(3)}$ and the Higgs Phenomenon

is no longer asserted in the literature that the mass of the neutrino is zero, so that the notion of a massless particle is being questioned. Nonetheless, the GWS point of view leads to the existence of massive bosons but simultaneously asserts by modelling that the photon mass is identically zero. Essentially, therefore, the Higgs phenomenon leads to a massive photon which is accepted in superconductors, but conventionally rejected in the vacuum, despite the fact that the same Higgs Lagrangian (and consequent Proca equation) is used in both superconductors and in the vacuum. Therefore the contemporary point of view appears to assert that photon mass is acceptable in superconductors but not acceptable in the vacuum, i.e., accepts the Proca equation in superconductors but rejects it in the vacuum. This is mutually incompatible - an equation of natural philosophy must be generally applicable, if the photon is asserted to be massive in superconductors it must be massive in the vacuum. If the Proca equation is valid in superconductors and if the photon mass is *a property of the photon*, and not something given to the photon by its interaction with superconducting material, then the Proca equation must be valid also in the vacuum.

2.1 CYCLICALLY SYMMETRIC EQUATIONS FOR FINITE PHOTON MASS

In this section the Einstein equation (1) is used to show that the non-Abelian equations (169) remain the same in structure in the presence of finite photon mass provided that the scalar amplitude $B^{(0)}$ is replaced by $B^{(0)} \exp(-\xi Z)$, where ξ is the *rest wave vector* given by $m_0 c/\hbar$ where m_0 is the intrinsic and irremovable photon mass. As argued in Vol. 1, ξ is a minute quantity, difficult to detect experimentally, but is, nonetheless, related directly to the well accepted notion of rest energy, $m_0 c^2$, for a particle with mass. Our derivation in this section is based on the well accepted fundamental axioms of quantum mechanics, which in Minkowski notation become

$$p_\mu = \hbar \kappa_\mu = -i\hbar \frac{\hat{\partial}}{\partial x_\mu}, \tag{176}$$

where $p_\mu = (\mathbf{p}, i(En/c))$ is the particulate energy-momentum four-vector and $\kappa_\mu = (\mathbf{\kappa}, i(\omega/c))$ the wave four-vector of radiation. Here ω is the angular frequency and $\mathbf{\kappa}$ the space part of the wave vector of matter waves. The matter wave is an electromagnetic wave if the mass of the particle is the photon mass. It is asserted conventionally that if the photon mass is zero, the axioms (176) still apply. In three dimensional notation the axioms of quantum mechanics are

$$En = \hbar\omega = i\hbar\frac{\hat{\partial}}{\partial t}, \qquad p = \hbar\kappa = -i\hbar\hat{\nabla}, \tag{177}$$

and historically, it was de Broglie's inference of matter waves that led to the development of wave mechanics in material such as atoms and molecules as well as in radiation.

Using the axioms (177) in the Einstein equation (1) leads to the following relation between the angular frequency, ω, and the wave-number κ,

$$\omega^2 = c^2\kappa^2 + \xi^2. \tag{178}$$

If the mass m_0 is identically zero, this relation becomes

$$\kappa = \frac{\omega}{c}, \tag{179}$$

which is the conventional relation between κ and ω for an electromagnetic wave propagating in the vacuum. Using the axioms (177), Eq. (178) is the Proca equation,

$$\Box A_\mu := \left(\frac{1}{c^2}\frac{\partial^2}{\partial t^2} - \nabla^2\right)A_\mu = -\xi^2 A_\mu, \tag{180}$$

if the wavefunction is identified with A_μ. The Proca equation is therefore intrinsically quantum mechanical in nature and as discussed in Chap. 1, is the same in structure as the Dirac equation. When $m_0 = 0$, and in the classical limit, the Proca equation (180) becomes the d'Alembert equation,

$$\Box A_\mu = 0, \tag{181}$$

which is an expression of the vacuum Maxwell equations. Defining $\kappa := \kappa' - i\kappa''$, Eq. (178) is a quadratic in κ:

$$\kappa^2 = \kappa'^2 - \kappa''^2 - 2i\kappa'\kappa'', \qquad Re(\kappa^2) = \frac{\omega^2}{c^2} - \xi^2 \tag{182}$$

46   Chapter 2. $B^{(3)}$ and the Higgs Phenomenon

and so κ is in general a complex quantity when the mass m_0 is non-zero. Specifically,

$$\kappa' := \frac{\omega}{c}, \quad \kappa'' := \xi, \qquad (183)$$

is compatible with Eq. (178). This allows the identification of the rest frequency of radiation of finite mass,

$$\hbar\omega_0 = m_0 c^2, \qquad (184)$$

which is de Broglie's Guiding Theorem (Eq. (1) of Vol. 1). In the particle interpretation, this means that the photon has a rest frame, a rest mass, and a rest energy. The *speed of light* c therefore becomes a postulated universal constant of special relativity, because a photon of finite mass does not propagate at c.

From Eq. (183) the phase of an electromagnetic plane wave becomes

$$\exp(i\phi) = \exp(i(\omega t - \kappa' Z)) \exp(-\xi Z), \qquad (185)$$

i.e., is the usual phase multiplied by the exponential $\exp(-\xi Z)$, a real quantity. The complex conjugate of the phase becomes

$$\exp((i\phi)^*) = \exp(-i(\omega t - \kappa' Z)) \exp(-\xi Z), \qquad (186)$$

and the cyclic relations (169) remain the same, provided that the scalar flux density magnitude, $B^{(0)}$, is replaced by $B^{(0)} e^{-\xi Z}$ as indicated already. Similarly, the scalar vacuum magnitude, $E^{(0)}$, of electric field strength is replaced by $E^{(0)} e^{-\xi Z}$. The spin field

$$\boldsymbol{B}^{(3)} = B^{(0)} e^{-\xi Z} \boldsymbol{e}^{(3)}, \quad \boldsymbol{e}^{(3)} = \boldsymbol{k}, \qquad (187)$$

is therefore a solution of the Proca equation, as are the wave fields

$$\boldsymbol{B}^{(1)} = \frac{B^{(0)}}{\sqrt{2}} e^{-\xi Z}(i\boldsymbol{i} + \boldsymbol{j})e^{+i(\omega t - \kappa' Z)},$$

(188)

$$\boldsymbol{B}^{(2)} = \frac{B^{(0)}}{\sqrt{2}} e^{-\xi Z}(-i\boldsymbol{i} + \boldsymbol{j})e^{-i(\omega t - \kappa' Z)},$$

Since these fields are solutions of Proca's equation, they are also generated by the Higgs phenomenon in the vacuum, a phenomenon which relates photon mass to spontaneous symmetry breaking, an essential ingredient of unified field theory. However, it is important to note that finite m_0 is not a *necessary* condition for the existence of the non-Abelian relations (169), i.e., $\boldsymbol{B}^{(3)}$ exists when $m_0 = 0$ identically and is a solution both of the d'Alembert and Proca equations in the vacuum. It is clear that $\boldsymbol{B}^{(3)}$ from the d'Alembert equation has no exponential decay, whereas $\boldsymbol{B}^{(3)}$ from the Proca equation decays exponentially according to Eq. (187), and so must ultimately vanish completely as Z approaches infinity.

2.2 LINK WITH THE HIGGS PHENOMENON

In the theory of fields and particles, the spontaneous breaking of gauge symmetry leads to the Proca equation, in the sense that massive bosons are predicted, and in unified field theory of the GWS type [16], are observed experimentally. The Higgs boson has a finite mass, but has not been observed experimentally. However, the realization that $\boldsymbol{B}^{(3)}$ exists in the non-trivial topology of the vacuum means that the photon must have three space polarizations, which is exactly what is inferred on the grounds that the photon may have mass, i.e., by the Proca equation. The latter is a result of the Higgs Lagrangian, as described by Ryder [16]. The experimental observation of $\boldsymbol{B}^{(3)}$ in the inverse Faraday effect, which is magnetization by light, lends support to the Higgs theory. Evidence for the Higgs phenomenon, and the Higgs boson, therefore becomes available through a combination of data from the inverse Faraday effect, which shows the existence of $\boldsymbol{B}^{(3)}$, and the evidence recently reviewed by Vigier [34] for photon mass, evidence which is assembled from several different sources. Standard tables no longer list the photon mass as identically zero, but as an upper limit inferred from experimental sources. Taken with the emergence of $\boldsymbol{B}^{(3)}$, therefore, the upper limit on photon mass is also an

Chapter 2. $B^{(3)}$ and the Higgs Phenomenon

upper limit on the mass of a Higgs boson — the massive photon. The existence of $B^{(3)}$ in electromagnetic theory has consequences in unified field theory which will be explored later in this book, and as we have seen, finite photon mass is not incompatible with gauge invariance provided that $A_\mu A_\mu = 0$. The field $B^{(3)}$ appears therefore to be compatible in every way with the Higgs phenomenon and non-trivial vacuum topology. In this section, the Proca equation is derived from the Higgs Lagrangian which leads to the massive photon, an inference which is accepted as experimentally plausible, i.e., no attempt is made to *model out* the photon mass as zero as in the GWS (or $SU(2) \otimes U(1)$) theory. The same Higgs Lagrangian is used in subsequent sections to show that $B^{(3)}$ is a topological string, i.e., a quantized magnetic flux density in one (Z) dimension. This string of magnetic flux density propagates in free space, i.e., through the non-trivial topology of the vacuum. The same Higgs Lagrangian gives rise to the equivalent of $B^{(3)}$ in type II superconductors, i.e., to Abrikosov lines. Although the vacuum is not a superconductor, the topological features of both vacuum and superconductor allow the existence of $B^{(3)}$. The Maxwellian point of view is therefore a limit in which the photon mass is identically zero.

It may be significant that the algebra (169) that links $B^{(3)}$ to $B^{(1)}$ and $B^{(2)}$ is non-Abelian, i.e., the cross products of fields are non-commutative. The same is true of the non-Abelian algebra between components of the vector potential, Eqs. (172), and the algebra,

$$E^{(1)} \times E^{(2)} = -E^{(0)}(iE^{(3)})^*, \qquad E^{(2)} \times (iE^{(3)}) = -E^{(0)} E^{(1)*},$$

$$(iE^{(3)}) \times E^{(1)} = -E^{(0)} E^{(2)*}, \tag{189}$$

between electric field components. Spontaneous symmetry breaking of a non-Abelian gauge theory leads to the GWS unified field theory [16], which is renormalizable. For this reason, massive bosons are acceptable in GWS and have been found experimentally. The existence of a massive photon is therefore not incompatible with renormalizability in quantum electrodynamics, and finite photon mass has been worked into GWS and SU(5), for example in papers by Huang [35]. It is therefore clear that $B^{(3)}$ is compatible with renormalizability in QED, and this subject is addressed later in this volume. Even in the absence of photon mass and in the absence of a Higgs field, the non-Abelian algebra (169), (172) and (189)

Link with the Higgs Phenomenon

remains valid, showing that the U(1) group of electromagnetism, the group of all numbers of the form

$$e^{i\alpha} = \cos\alpha + i\sin\alpha, \qquad (190)$$

should always be viewed as including the case $\alpha = 0$. If α is identified with the electromagnetic phase, then $\alpha = 0$ implies

$$\omega t - \kappa \cdot r = 2\pi n, \qquad (191)$$

where n is an integer. This is precisely the string condition [16]

$$e\Phi = -2\pi n, \qquad (192)$$

where Φ is the magnetic flux associated with the vortex line, so that the string-like magnetic flux is identified with

$$\Phi := \frac{1}{e}(\kappa \cdot r - \omega t). \qquad (193)$$

This finding is consistent, furthermore, with the fact that non-linear, classical field theories in general produce solitons [16]. The non-Abelian relations (169), (172) and (189) are also non-linear, and significantly, $B^{(3)}$ is a solution which is a stable configuration with a well defined energy which is nowhere singular. The field $B^{(3)}$ therefore has all the characteristics of a soliton solution, being the expectation value of the photomagneton of quantized magnetic flux. In recent years [16] non-Abelian gauge theories have predicted the existence of vortices, magnetic monopoles of the 't Hooft-Polyakov variety [36], and instantons, which are soliton solutions to the gauge field equations in two space dimensions (i.e., a string in three dimensional cylindrical space). In the same way that the stability of soliton solutions in non-linear field theories is a consequence of topology, the stability of $B^{(3)}$ in free space is a direct consequence of the existence of the complex conjugate wave fields $B^{(1)}$ and $B^{(2)}$. In other words $B^{(3)}$ is linked topologically with $B^{(1)}$ and $B^{(2)}$, all three fields being proportional to rotation generators of the Poincaré group of space-time. The topology of the U(1) group ensures the stability of

Chapter 2. $B^{(3)}$ and the Higgs Phenomenon

the $B^{(3)}$ vortex in the vacuum, a vortex line that carries magnetic flux of finite energy.

The structure of the Proca equation of field theory can be deduced as follows from the Klein-Gordon equations for a complex, scalar field, denoted by

$$(\Box + \xi_\phi^2)\phi = 0, \tag{194a}$$

$$(\Box + \xi_\phi^2)\phi^* = 0, \tag{194b}$$

where $\xi_\phi := (m_\phi c)/\hbar$, in S.I. units, where $\hbar$ and c are not suppressed. In writing the Klein-Gordon equations in this way, quantization has already been assumed, because of the presence of the Dirac constant, $\hbar$. In a classical approach, as described by Ryder [16], the scalar field ϕ is not a single particle wave function before canonical quantization, but can be regarded as a generalized coordinate, i.e., ϕ replaces a coordinate x and the time t of the function x(t) is generalized to x_μ. In this view, the two Klein-Gordon equations are classical wave equations in which the parameter m_ϕ is not immediately identifiable as the mass of a point particle. Canonical quantization of the complex Klein-Gordon field produces, interestingly, particles and anti-particles with the same mass but opposite charge, spin being unconsidered because we have a scalar field. If it is possible, as we have asserted, to produce a Proca equation from the two Klein-Gordon equations (194a) and (194b), then if these are classical, so must be the Proca equation itself. Canonical quantization of this classical Proca equation then proceeds satisfactorily, because the particle has *three* space-like polarizations [16]. If the particle is a photon, then introducing the mass parameter m_ϕ means introducing a third space-like polarization, *which is precisely what is indicated by the non-Abelian equations (169), (172) and (189)*.

The two Klein-Gordon equations (194a) and (194b) can be obtained from the Lagrange equation of motion,

$$\frac{\partial \mathscr{L}_\phi}{\partial \phi} - \frac{\partial}{\partial x_\mu}\left(\frac{\partial \mathscr{L}_\phi}{\partial\left(\frac{\partial \phi}{\partial x_\mu}\right)}\right) = 0, \tag{195}$$

using the Lagrangian,

Link with the Higgs Phenomenon

$$\mathcal{L}_\phi = \frac{\hbar^2}{2m_\phi} \frac{\partial \phi}{\partial x_\mu} \frac{\partial \phi^*}{\partial x_\mu} - \frac{1}{2} m_\phi c^2 \phi \phi^*, \tag{196}$$

in S.I. units. In this classical approach m_ϕ is a parameter with the units of mass, a parameter that is generated by the wave equations (194a) and (194b), which originate as described in Chap. 1 of this volume, from the Einstein equation of special relativity. In that classical particle equation, m_ϕ is the particle mass.

Applying the contemporary principles of gauge invariance to the Lagrangian (196) results in the introduction of a four-vector A_μ, which has all the properties of the potential four vector of electromagnetism [37]. In order to preserve gauge invariance, and the principles of special relativity, the Lagrangian (196) must be modified to

$$\mathcal{L} = \frac{1}{2m_\phi}\left(\hbar \frac{\partial \phi}{\partial x_\mu} + ieA_\mu \phi\right)\left(\hbar \frac{\partial \phi^*}{\partial x_\mu} - ieA_\mu \phi^*\right)$$
$$- \frac{1}{2} m_\phi c^2 \phi \phi^* - \frac{1}{4} \epsilon_0 F_{\mu\nu} F_{\mu\nu}^*, \tag{197}$$

where ϵ_0 is the free space permittivity, $F_{\mu\nu}$ the field four-tensor of electromagnetism, and e the charge on the electron. Using the quantum prescription (176), the Lagrangian (197) becomes the classical

$$\mathcal{L} = \frac{1}{2m_\phi}(ip_\mu \phi + ieA_\mu \phi)(-ip_\mu \phi^* - ieA_\mu \phi^*)$$
$$- \frac{1}{2} m_\phi c^2 \phi \phi^* - \frac{1}{4} \epsilon_0 F_{\mu\nu} F_{\mu\nu}^*, \tag{198}$$

in which the Dirac constant $\hbar$ has disappeared. Note that nothing has yet been mentioned of spontaneous symmetry breaking, which will be interwoven into these considerations at a later stage. The mass parameter m_ϕ appearing in Eq. (198) originates in the complex Klein-Gordon field, and not the electromagnetic field, which introduced itself through A_μ as a consequence of special relativity expressed through gauge invariance. Readers are referred to Ryder [16] for more details of this process. The Lagrangian (196) of the complex Klein-Gordon field is not compatible with special relativity because, essentially speaking, it allows action at a distance. The Lagrangian (198) is made compatible with special relativity by the introduction of A_μ through the covariant derivative

$$D_\mu := \frac{\partial}{\partial x_\mu} + \frac{ieA_\mu}{\hbar}, \qquad (199)$$

which shows that eA_μ has the same S.I. units as $-i\hbar(\partial/\partial x_\mu)$. Electromagnetism is therefore seen to be a property of space-time itself. This deduction was repeatedly reinforced in Vol. 1 through the fact that $B^{(3)}$ is directly proportional to a rotation generator of the Poincaré group of space-time, and is therefore physically meaningful in the vacuum.

The Lagrange equation

$$\frac{\partial \mathcal{L}}{\partial A_\mu} - \frac{\partial}{\partial x_\nu}\left(\frac{\partial \mathcal{L}}{\partial\left(\frac{\partial A_\mu}{\partial x_\nu}\right)}\right) = 0, \qquad (200)$$

with the Lagrangian (197) leads to the *Proca equation*,

$$\left(\Box - \frac{e^2 \phi \phi^*}{m_\phi}\right) A_\mu = -eJ_\mu, \qquad (201)$$

where the current J_μ is defined as in Chap. 1 by

$$J_\mu = \frac{\hbar}{2m_\phi}\left(\phi^* \frac{\partial \phi}{\partial x_\mu} - \phi \frac{\partial \phi^*}{\partial x_\mu}\right). \qquad (202)$$

Note that there is a term in $A_\mu A_\mu$ in the Lagrangian (197), which is identifiable as a mass term. This makes its appearance on the left hand side of the Proca equation (201) and is consistent with gauge invariance. The right hand side describes the interaction of e with a current J_μ which is premultiplied by the Dirac constant $\hbar$. In classical physics, however, this is zero, so the right hand side of the Proca equation (201) has no classical equivalent and vanishes in classical physics. The left hand side remains the same if m_ϕ can be regarded as a classical mass. This result is equivalent to asserting that the complex Klein-Gordon field has no linear momentum, which is represented entirely by the electromagnetic field in the entirely *classical* Lagrangian,

$$\mathcal{L}_0 = \frac{e^2}{2m_\phi} A_\mu A_\mu \phi \phi^* - \frac{1}{2} m_\phi c^2 \phi \phi^* - \frac{1}{4} \epsilon_0 F_{\mu\nu} F^*_{\mu\nu}, \qquad (203)$$

Link with the Higgs Phenomenon

equivalent to the classical Proca equation

$$\left(\Box - \frac{e^2\phi\phi^*}{m_\phi}\right)A_\mu = 0. \tag{204}$$

This is recognizable as the *vacuum Proca equation* which obtains conventionally when the source of the electromagnetic field is infinitely distant. The mass term in this equation is obtained from the product

$$\xi^2 := -\frac{e^2\phi\phi^*}{m_\phi}, \tag{205}$$

and has been given up to the electromagnetic field by the complex Klein-Gordon field. In quantum physics, the Dirac constant $\hbar$ is not zero, and so the current term on the right hand side of Eq. (201) is restored. This can be regarded as the source of the electromagnetic field, a source which resides in a charged, complex scalar field. Therefore the Proca equation (201) describes an interaction process in which linear momentum is conserved. In this view, therefore, quantum physics asserts that the source of electromagnetism can never be infinitely distant, as in classical physics, because in quantum physics, $\hbar$ is non-zero. The charge e is not quantized in this view, and is introduced through the product eA_μ.

It is inferred that if the electromagnetic field's source is the complex Klein-Gordon field, then the latter gives up mass to the former, a mass term described by Eq. (205).

In this view, therefore, the classical electromagnetic field can never be without mass, because it would have no source. As in Chap. 1, the Proca equation,

$$(\Box + \xi^2)A_\mu = 0, \tag{206}$$

is obtainable directly from the Einstein equation (1) by regarding A_μ as a wave function and using the quantum prescription (176). Equation (206) is therefore a wave equation equivalent to a free photon. Identifying the two mass terms gives

$$\xi^2 = \left(\frac{m_0 c}{\hbar}\right)^2 = -\frac{e^2\phi\phi^*}{m_\phi}, \tag{207}$$

where m_0 is the intrinsic, irremovable, mass of the free photon.

The current term,

$$J_\mu^{(eff)} := \left(\frac{e\phi\phi^*}{m_\phi}\right)A_\mu, \qquad (208)$$

is a London equation in the vacuum, and if a vacuum resistance R can be defined, and if Ohm's Law can be assumed to apply,

$$E_\mu = R\left(\frac{e\phi\phi^*}{m_\phi}\right)A_\mu, \qquad (209)$$

showing that there is an electric field E_μ generated directly by A_μ through the Proca equation. This is discussed by Moles and Vigier [38]. The existence of the field $\boldsymbol{B}^{(3)}$ follows directly from Ampère's equation,

$$\nabla \times \boldsymbol{B}^{(3)} = \boldsymbol{J}^{(eff)}, \qquad (210)$$

which gives

$$\nabla^2 \boldsymbol{B}^{(3)} = \xi^2 \boldsymbol{B}^{(3)}. \qquad (211)$$

The solution of this equation is

$$\boldsymbol{B}^{(3)} = B^{(0)} e^{-\xi z} \boldsymbol{k}, \qquad (212)$$

which identifies the rest wave-number of Eq. (183) of this chapter as

Link with the Higgs Phenomenon

$$\xi := \kappa_0 = \left(-\frac{e^2 \phi \phi^*}{m_\phi}\right)^{\frac{1}{2}}. \tag{213}$$

Equation (212) is of course consistent with the Proca equation (206).

2.2.1 SPONTANEOUS SYMMETRY BREAKING

Spontaneous symmetry breaking is characterized [39] by the equation

$$\phi \phi^* := |\phi|^2 = -\frac{M^2}{2\lambda}, \tag{214}$$

where M is a *parameter*. In this view, introduced by Higgs [40] and others, the vacuum becomes the state of lowest potential energy, the minimum value of

$$V(\phi) = M^2 \phi^2 + \lambda \phi^4, \tag{215}$$

given by Eq. (214). The vacuum is no longer necessarily the state in which the field is absent. Therefore the field has a vacuum expectation value, given by

$$\phi \phi^* = |\langle 0|\phi|0\rangle|^2, \tag{216}$$

vacuum eigenstates being denoted by $|0\rangle$. From the equation

$$\phi_{min} := a = \pm\left(-\frac{M^2}{2\lambda}\right)^{\frac{1}{2}}, \tag{217}$$

mass becomes something that ensures the minimization of energy — an embodiment of a variational principle. Therefore spontaneous symmetry breaking (SSB) adds a background energy to the universe, and is characterized by the addition of the term $\lambda \phi^4$ to the Lagrangian (203). From Eq. (213), the photon mass m_0 can be expressed directly in terms of the mass parameter M, and in this sense, the photon *picks up mass* from the vacuum as it propagates. This notion is furthermore consistent with the recent inference [41] of *vacuum friction*, which explains how light intensity can be lost exponentially as the light beam propagates through the vacuum. Since light intensity (watts m^{-2}) is defined in S.I. units by

$$I_0 = \epsilon_0 c E^{(0)2} = \frac{1}{\mu_0} c B^{(0)2}, \tag{218}$$

where μ_0 is the vacuum permeability and ϵ_0 the vacuum permittivity, then a massive photon implies that intensity decays exponentially along the propagation axis (Z). If so, then electromagnetic energy density, defined by

$$U_0 = \frac{1}{2}\left(\epsilon_0 E^{(0)2} + \frac{1}{\mu_0} B^{(0)2}\right), \tag{219}$$

is lost as light propagates (Tolman's *tired light* [41]), and the range of electromagnetic waves becomes finite, even if they are propagating through a vacuum. These inferences require a vacuum resistance R, as in Eq. (209), or *vacuum friction*.

The symmetry broken Higgs Lagrangian has been shown in recent years to produce a soliton solution, a vortex line of quantized magnetic flux which in type two superconductors is an Abrikosov line [42]. This vortex is stable in two or more dimensions [16] if and only if there is also present a gauge field such as an electromagnetic field represented by A_μ. The vortex line appears in two dimensional space, or *three dimensional* space with cylindrical symmetry (both characterized by U(1), the group of numbers $e^{i\alpha}$), and is therefore identifiable directly with **B**$^{(3)}$ in the vacuum. Thus, the field **B**$^{(3)}$ emerges from *the Klein-Gordon equation of an electron in A_μ*, as deduced in Chap. 1 for the Dirac equation of e in A_μ. For reasons discussed there, the Klein-Gordon equation must be replaced by the Dirac equation in order to produce a physical probability density and in order to produce a correct description of anti-particles, but each component of the Dirac equation must also satisfy a Klein-Gordon equation. As discussed by Ryder [16], the Dirac equation with considerations of SSB leads directly to the GWS unification of the electromagnetic and weak fields. In this view therefore the association of **B**$^{(3)}$ with photon mass becomes inevitable, so that the well known experimental observation [43] of **B**$^{(3)}$ in the inverse Faraday effect becomes indirect but persuasive evidence for the existence of photon mass and therefore for a Higgs boson.

The close relation between SSB and the non-Abelian algebra (169) that defines **B**$^{(3)}$ in the group O(3) of rotations

Link with the Higgs Phenomenon

in 3-D space can be developed as follows. In the conventional view of vacuum electrodynamics [44-46], the electromagnetic energy density is defined by the transverse, wave fields, and by these only,

$$U = \frac{1}{2}\left(\frac{1}{\mu_0}(\boldsymbol{B}^{(1)} \cdot \boldsymbol{B}^{(1)*} + \boldsymbol{B}^{(2)} \cdot \boldsymbol{B}^{(2)*})\right. \tag{220}$$

$$\left. + \epsilon_0(\boldsymbol{E}^{(1)} \cdot \boldsymbol{E}^{(1)*} + \boldsymbol{E}^{(2)} \cdot \boldsymbol{E}^{(2)*})\right),$$

in the circular basis [47] of Vol. 1,

$$\boldsymbol{e}^{(1)} = \boldsymbol{e}^{(2)*} = \frac{1}{\sqrt{2}}(\boldsymbol{1} - i\boldsymbol{j}), \qquad \boldsymbol{e}^{(3)} = \boldsymbol{k}. \tag{221}$$

The energy (220) can be denoted in Minkowski notation by the product of the electromagnetic four-tensor $F_{\mu\nu}$ with the complex conjugate tensor $F_{\mu\nu}^*$. In the circular basis these tensors must be defined by

$$F_{\mu\nu} := \begin{bmatrix} 0 & cB^{(3)} & -cB^{(2)} & -iE^{(1)} \\ -cB^{(3)} & 0 & cB^{(1)} & -iE^{(2)} \\ cB^{(2)} & -cB^{(1)} & 0 & -iE^{(3)} \\ iE^{(1)} & iE^{(2)} & iE^{(3)} & 0 \end{bmatrix},$$

$$F_{\mu\nu}^* := \begin{bmatrix} 0 & cB^{(3)*} & -cB^{(2)*} & iE^{(1)*} \\ -cB^{(3)*} & 0 & cB^{(1)*} & iE^{(2)*} \\ cB^{(2)*} & -cB^{(1)*} & 0 & iE^{(3)*} \\ -iE^{(1)*} & -iE^{(2)*} & -iE^{(3)*} & 0 \end{bmatrix},$$

(222)

in order to obtain a positive U from the tensor product $\epsilon_0 F_{\mu\nu} F_{\mu\nu}^*$. From the principle of gauge invariance of the second kind [16], the term U must be subtracted from the complex Klein-Gordon Lagrangian density to maintain consistency with special relativity. Therefore the term equivalent to the energy U in the gauge invariant Lagrangian density is

$$\mathcal{L}_{no\,mass} - \frac{1}{4}\epsilon_0 F_{\mu\nu} F_{\mu\nu}^*. \tag{223}$$

The non-Abelian algebra (169) and (189), however, indicates that this must contain terms due to $\boldsymbol{B}^{(3)}$ and $-i\boldsymbol{E}^{(3)}/c$ in the vacuum, i.e., real and imaginary field terms that are phase free and which require for their definition the involvement of the third polarization space axis (3), defined by the unit vector

$$e^{(3)} = k, \qquad (224)$$

which is in turn identified precisely in the circular basis (221) with the cartesian unit vector k in the propagation axis Z of the light beam. In the particle interpretation therefore, *the photon has three space dimensions*. This is precisely the outcome of SSB, which in the presence of electromagnetism produces a vortex line, $B^{(3)}$ in the vacuum, which in type II superconductors becomes an Abrikosov line. Furthermore, the structure of the non-Abelian algebra (169) is maintained exactly for the electromagnetic field with mass provided $B^{(0)}$ is replaced by $B^{(0)}e^{-\xi z}$.

The existence of $B^{(3)}$ and its dual (Vol. 1) $-iE^{(3)}/c$ in the vacuum therefore leads to the replacement of (223) by

$$\mathcal{L}_{mass} = -\frac{1}{4}\epsilon_0 F_{\mu\nu} F^*_{\mu\nu} e^{-2\xi z}$$

$$= -\frac{1}{2}\left(\frac{1}{\mu_0}(B^{(1)} \cdot B^{(1)*} + B^{(2)} \cdot B^{(2)*} + B^{(3)} \cdot B^{(3)*})\right. \qquad (225)$$

$$\left. + \epsilon_0(E^{(1)} \cdot E^{(1)*} + E^{(2)} \cdot E^{(2)*} + E^{(3)} \cdot E^{(3)*})\right)e^{-2\xi z}.$$

The exponential in Eq. (225) is represented by

$$e^{-2\xi z} \sim 1 - 2\xi z + 2\xi^2 z^2 + \ldots, \qquad (226)$$

and the difference (*mass correction*) between the Lagrangian densities in the presence and absence of electromagnetic mass becomes

$$\Delta\mathcal{L} = \mathcal{L}_{no\,mass} - \mathcal{L}_{mass} = -\frac{1}{2}\epsilon_0 F_{\mu\nu} F^*_{\mu\nu}(\xi z - \xi^2 z^2), \qquad (227)$$

an energy difference with the same structure as the SSB energy introduced by Higgs, Eq. (215), and with a minimum defined by

Link with the Higgs Phenomenon

$$\frac{\partial(\Delta\mathcal{L})}{\partial Z} = -\frac{1}{2}\epsilon_0 F_{\mu\nu}F_{\mu\nu}^*(\xi - 2\xi^2 Z) = 0. \tag{228}$$

The field $B^{(3)}$ in the vacuum can therefore be considered in terms of spontaneous symmetry breaking in this way. Qualitatively, $B^{(3)}$ has a rod-like symmetry which can be spontaneously broken to produce $B^{(1)}$ or $B^{(2)}$ in the plane perpendicular to $B^{(3)}$. The latter is independent of phase, and so the symmetry breaking can produce any value of the phase without affecting the magnitude or direction of $B^{(3)}$, i.e., the fields $B^{(1)}$ and $B^{(2)}$ can be oriented *randomly* with respect to $B^{(3)}$ and still produce the same value of $B^{(3)}$. This process is directly analogous with the spontaneous bending of a rod in any direction as described in Ryder [47].

Continuing the analogy between SSB, intrinsic photon mass, m_0, and $B^{(3)}$, it is seen that $-\xi Z$ plays the role of $M^2\phi^2$; $\xi^2 Z^2$ plays the role of $\lambda\phi^4$; and $\Delta\mathcal{L}$ plays the role of V. In the absence of intrinsic photon mass, the usual vacuum state occurs at $\Delta\mathcal{L} = 0$ (i.e., $Z = 0$), in which case

$$\mathcal{L}_{no\,mass} = \mathcal{L}_{mass}. \tag{229}$$

The symmetry broken vacuum state, on the other hand, is given by a minimum in the *difference* $\Delta\mathcal{L}$, a minimum defined by

$$\frac{\partial(\Delta\mathcal{L})}{\partial Z} = 0, \tag{230}$$

so that $\xi Z = 1/2$. At this minimum point

$$(\Delta\mathcal{L})_{min} = -\frac{1}{4}\epsilon_0 F_{\mu\nu}F_{\mu\nu}^*. \tag{231}$$

The SB vacuum state, which no longer indicates the *absence* of the electromagnetic field tensor $F_{\mu\nu}$, but is a *minimum* of the electromagnetic field between vacuum eigenstates, is therefore displaced by Eq. (231) from the usual vacuum state. If, following the Higgs method [16], we define the minimum of the electromagnetic Lagrangian to occur at $Z = 1/(2\xi)$, at which the numerical value of the Lagrangian is set to zero, then at $Z = 0$, the numerical value of the same Lagrangian function must be $-(1/4)\epsilon_0 F_{\mu\nu}F_{\mu\nu}^*$, which represents a local maximum at $Z = 0$. This conclusion can be checked through the fact that the value of

$$\mathcal{L}_{mass} := -\frac{1}{4}\epsilon_0 F_{\mu\nu}F^*_{\mu\nu}e^{-2\xi Z} \qquad (232)$$

at $Z = 0$ is of course $-(1/4)\epsilon_0 F_{\mu\nu}F^*_{\mu\nu}$. If the exponential is not approximated by the first two terms of a Maclaurin series, as in Eq. (226), its minimum occurs at $Z \to \infty$, and the condition $\xi Z = 1/2$ represents a characteristic correlation distance of the exponential decay of the Lagrangian function with photon mass. The Lagrangian decays to zero in this case only as Z approaches infinity, in which limit gauge invariance (and special relativity) can no longer be maintained. However, since the radius of the universe is finite, the condition $Z \to \infty$ is unphysical, so that for all physical situations, gauge invariance is maintained by the presence of electromagnetism.

If the exponential is approximated as in Eq. (226), however, the approximation, a Maclaurin series, is valid only for $\xi Z \le 1$, beyond which it is mathematically invalidated. Therefore inferences based on Eq. (226) are valid if and only if $\xi Z \le 1$, which fixes the range of validity. The minimum in Eq. (231) occurs by definition only within the range of validity of the Maclaurin series. For all practical purposes ξZ is always much less than one because the numerical value of ξ is less than of the order (Vol. 1) 10^{-52} m^{-1}. Therefore the Maclaurin series is approximated excellently by its first two terms as in Eq. (226) in all laboratory experiments in physical optics. In cosmology, however, Z can approach the known radius of the universe in order of magnitude. In summary, the appearance of an exponential decay in the Lagrangian (232) is analogous with spontaneous symmetry breaking, i.e., the existence of finite intrinsic photon mass, m_0, implies that the vacuum state of electromagnetism is a minimum of the electromagnetic field, and not the state in which the field is absent. In the following section, it is shown that the conventional symmetry broken Higgs Lagrangian leads directly to the inference that $B^{(3)}$ is a vortex line, or soliton solution, of the non-linear complex Klein-Gordon equations in the presence of electromagnetism, the latter being an inevitable consequence of gauge invariance of the second kind. Thus, $B^{(3)}$ is also an inevitable consequence of gauge invariance.

2.2.2 $B^{(3)}$ AS A VORTEX LINE IN THE VACUUM

In Chap. 1 it was shown that the vacuum $B^{(3)}$ is a consequence of the standard Dirac equation of e in A_μ. Since Klein-Gordon equations describe the evolution of individual spinor components and also of scalar components of A_μ, it follows that $B^{(3)}$ must also emerge from the Klein-Gordon equation of e interacting with the electromagnetic field, using the same minimal prescription introduced through the same covariant derivative, Eq. (120). In proving this result in this section, it is also demonstrated that $B^{(3)}$ is a vortex line in the vacuum, a soliton solution of the appropriate non-linear field equations. Since electromagnetism itself is a consequence [16] of the need to keep the complex Klein-Gordon Lagrangian invariant under gauge transformation, the demonstration in this section proves that the vacuum vortex line $B^{(3)}$ is an inevitable consequence of gauge invariance in the complex Klein-Gordon equation of field theory. It is important to note, however, that this conclusion holds whether or not the photon is regarded as having intrinsic mass, m_0, and is valid in the presence or absence of spontaneous symmetry breaking. However, as argued already, it seems overwhelmingly probable that $B^{(3)}$ *indicates* the existence of non-zero m_0, which as shown in Section 2.2.1, is mathematically analogous with spontaneous symmetry breaking. The latter is a key ingredient of unified field theory. If $B^{(3)}$ is asserted to be zero, while $B^{(1)}$ and $B^{(2)}$ are maintained to be non-zero, then field theory in general is invalidated. The belated recognition of $B^{(3)}$ therefore reinforces field theory as currently understood, and the inference that $B^{(3)}$ is non-zero in the vacuum is a major step forward in contemporary understanding. Experimental investigation of its characteristic square root power density dependence would therefore be of key importance.

The development in this Section is based on the Higgs Lagrangian (197) with the addition of the symmetry breaking term $-\lambda(\phi\phi^*)^2$. This Lagrangian in Eq. (200) produces Eq. (201), a Proca field equation. If we set $\phi = \phi^* = 0$ in Eq. (201) we recover the vacuum d'Alembert equation $\Box A_\mu = 0$ of which $B^{(3)}$ is a solution (Vol. 1). In the quantized interpretation the d'Alembert equation is equivalent to the equation of motion of a *free* particle, the photon free of the Klein-Gordon field. If this latter is taken to be the field of an electron interacting with the electromagnetic field represented by A_μ, then the interaction equations are obtained from the symmetry broken Lagrangian (197) in the two Lagrange equations

$$\frac{\partial \mathscr{L}}{\partial \phi} = D_\mu\left(\frac{\partial \mathscr{L}}{\partial(D_\mu\phi)}\right), \qquad (233)$$

$$\frac{\partial \mathscr{L}}{\partial A_\mu} = \partial_\nu\left(\frac{\partial \mathscr{L}}{\partial(\partial_\nu A_\mu)}\right), \qquad (234)$$

giving two non-linear equations which must be solved simultaneously [48],

$$\frac{1}{2m_\phi}D_\mu(D_\mu\phi^*) = -\frac{1}{2}m_\phi c^2\phi^* - 2\lambda\phi^*(\phi\phi^*), \qquad (235)$$

$$\left(\Box - e^2\frac{\phi\phi^*}{m_\phi}\right)A_\mu = -eJ_\mu. \qquad (236)$$

The covariant derivative D_μ plays the same role in these equations as in the Dirac equation (121), which was solved for e in the presence of A_μ, but without a symmetry breaking term. In Eqs. (235) and (236) there is a complicated interdependence of the Klein-Gordon and electromagnetic field due essentially to the principle of gauge invariance of the second kind, i.e., to the conservation of local as well as global symmetry. In addition, the vacuum is represented by

$$|\phi|_{vac} = a = \left(-\frac{M^2}{2\lambda}\right)^{\frac{1}{2}}, \qquad (237)$$

and the field ϕ is parameterized [16] as

$$\phi = \chi(r)e^{in\theta}, \qquad (238)$$

where

$$\chi(r) \to 0 \text{ as } r \to 0,$$
$$\chi(r) \to a \text{ as } r \to \infty. \qquad (239)$$

This means that the limit $\phi \to 0$ as $r \to 0$ is associated with finite energy, and with a finite $B^{(3)}$ field which is a vortex line of quantized magnetic flux, a stable soliton solution of the coupled non-linear field equations (235) and (236). Therefore, as anticipated, a stable $\boldsymbol{B}^{(3)}$ emerges from the Klein-Gordon equation of motion as well as from the Dirac equation of motion of e in A_μ. For further details the reader is referred to the original paper [48] or to the description given by Ryder [16] in reduced units. By a suitable parameterization of A_μ in polar coordinates, the field $\boldsymbol{B}^{(3)}$ is given in this view a *radial* dependence, i.e., a dependence on the radial coordinate r. In the limit $r \to 0$, $\boldsymbol{B}^{(3)}$ from Eqs. (235) and (236) is finite, and can be identified with $\boldsymbol{B}^{(3)}$ of the non-Abelian algebra (169). As $r \to \infty$, $\boldsymbol{B}^{(3)}$ disappears, i.e., it has a finite radius. Note carefully, however, that there is no dependence of $\boldsymbol{B}^{(3)}$ on Z, the propagation axis perpendicular to $\boldsymbol{r}$. Therefore $\boldsymbol{B}^{(3)}$ exists in the free photon as an infinitely narrow flux vortex line. The presence of the Klein-Gordon field gives $\boldsymbol{B}^{(3)}$ a finite radius through the use of spontaneous symmetry breaking.

Chapter 3. $B^{(3)}$ and Non-Abelian Gauge Geometry

The geometry of gauge fields plays a central role in contemporary field theory. Electromagnetism is conventionally asserted to be the U(1) sector of unified field theory, where U(1) is the group of numbers of the form $e^{i\alpha}$, or the group (O(2)) of rotations in a plane. Unified GWS field theory, for example, is then built on a product group such as SU(2) ⊗ U(1). However, equations such as (169), (172) and (189) are non-Abelian because of the presence of vector cross products, while the U(1) group is Abelian. In Vol. 1, it has been shown that $B^{(1)}$, $B^{(2)}$ and $B^{(3)}$ form a Lie algebra of the *non-Abelian* group of rotation matrices in space, O(3); matrices which are isomorphic with the Pauli spinors of SU(2), another non-Abelian group. In this chapter, a potential model for $B^{(1)}$, $B^{(2)}$ and $B^{(3)}$ is constructed from the general theory of gauge field geometry [16], *and the momentous conclusion reached that electromagnetism is a non-Abelian gauge field, described by the group O(3) in three dimensional space, rather than the group O(2) := U(1) of rotations in a plane.* This means that all field theories based on the assertion that U(1) is the sector of electromagnetism must be fundamentally modified. For example GWS theory must be constructed from a direct product group SU(2) ⊗ O(3) rather than SU(2) ⊗ U(1). Essentially speaking, our current appreciation of unified field theory is incomplete because the role of $B^{(3)}$ in the electromagnetic sector has not been realized. Relations such as (169) become Abelian if and only if

$$B^{(1)} \times B^{(2)} = B^{(2)} \times B^{(1)} =? \; 0, \qquad (240)$$

an assertion which is contradicted *experimentally* in data which have been available for some thirty years, for example in the inverse and optical Faraday effects, discussed in detail in Chap. 7 of Vol. 1, and in light shift data in atomic spectra [49], which have been available since about 1960 [50].

Chapter 3. $B^{(3)}$ and Non-Abelian Gauge Geometry

The general theory of gauge field geometry [16] is used in this chapter to infer that the electromagnetic four-tensor is described for electromagnetism as the O(3) sector, rather than the U(1) sector, by the non-linear

$$G_{\mu\nu} = \frac{\partial A_\nu}{\partial x_\mu} - \frac{\partial A_\mu}{\partial x_\nu} - i\frac{e}{\hbar}[A_\mu, A_\nu], \qquad (241)$$

where S.I. units have been used as throughout these volumes, so $\hbar$ has not been suppressed. From Eq. (241) we obtain

$$B_Z^{(3)} = -i\frac{e}{\hbar}[A_X^{(1)}, A_Y^{(2)}], \qquad (242)$$

showing that $B^{(3)}$ is defined in the non-Abelian O(3) electromagnetic sector by a commutator, which is the quantized version of the *conjugate product*,

$$\boldsymbol{B}^{(3)} = -i\frac{\kappa^2}{B^{(0)}}\boldsymbol{A}^{(1)} \times \boldsymbol{A}^{(2)}. \qquad (243)$$

This type of conjugate product has been discussed in detail in Chap. 1 of this volume, in the context of the Dirac equation. The indices (1), (2) and (3) appearing in Eq. (242) play the role of *isospin indices* in the well known Yang-Mills formalism [16] of non-Abelian field theory.

The extension of the group symmetry of electromagnetism from O(2) (or U(1)) to O(3) is a direct consequence of the *experimental* existence [51] of the conjugate product $\boldsymbol{B}^{(1)} \times \boldsymbol{B}^{(2)}$ and therefore of $\boldsymbol{B}^{(3)}$. The experimental presence of $\boldsymbol{B}^{(3)}$ in the vacuum means the presence of a *physical third axis*, an axis which as we have seen in Chap. 2 is already recognized in some contexts as a soliton solution, a quantized flux line, which in type II superconductors is an Abrikosov line. The group O(2) is Abelian, the group O(3) is non-Abelian, and the new *non*-Abelian dimension of electromagnetism in the vacuum is strongly indicative of the existence both of photon mass and of magnetic monopoles of the type first proposed by 't Hooft [52] and Polyakov [53]. Further experimental and theoretical work on $\boldsymbol{B}^{(3)}$ therefore becomes centrally important, because it is the physical key to the philosophical transition from O(2) to O(3) in the electromagnetic sector of contemporary unified field theory. It is particularly important in this context to experiment on the magnetization of electron plasma by circularly polarized electromagnetic radiation, as described in Chap. 12 of Vol. 1.

Although $B^{(3)}$ has been indirectly recognized [16] as a soliton solution, its link with $B^{(1)}$ and $B^{(2)}$, the cyclically symmetric vacuum equations (169), was recognized only in 1992 [1-12], *and it is this link that shows conclusively that electromagnetism in the vacuum is described by O(3) and its rotation generators, as developed in detail in Vol. 1.* In previous chapters of Vols. 1 and 2 it has been demonstrated rigorously that $B^{(3)}$ emerges from all the standard equations of one electron in the electromagnetic field: the classical relativistic Hamilton-Jacobi equation (Chap. 12 of Vol. 1); the Dirac equation (Chap. 1 of this volume); the complex Klein-Gordon equation (Chap. 2 of this volume). For the free field, $B^{(3)}$ is a solution of the d'Alembert and Proca equations. There is no further reasonable doubt therefore of the existence of the non-Abelian relations (169), (172) and (189) in the vacuum, relations which signal the emergence of electromagnetism in the vacuum as a non-Abelian gauge theory. The indices (1), (2) and (3) of the basis (221) now emerge as *isospin indices*. The familiar definition of the electromagnetic four-tensor in U(1) (or O(2))

$$F_{\mu\nu} = \frac{\partial A_\nu}{\partial x_\mu} - \frac{\partial A_\mu}{\partial x_\nu}, \qquad (244)$$

is generalized within the rigorous [16] mathematical theory of gauge geometry to

$$G_{\mu\nu}^{(1)*} = \partial_\mu A_\nu^{(1)*} - \partial_\nu A_\mu^{(1)*} - i\frac{e}{\hbar}\left[A_\mu^{(2)}, A_\nu^{(3)}\right],$$

$$G_{\mu\nu}^{(2)*} = \partial_\mu A_\nu^{(2)*} - \partial_\nu A_\mu^{(2)*} - i\frac{e}{\hbar}\left[A_\mu^{(3)}, A_\nu^{(1)}\right], \qquad (245)$$

$$G_{\mu\nu}^{(3)*} = \partial_\mu A_\nu^{(3)*} - \partial_\nu A_\mu^{(3)*} - i\frac{e}{\hbar}\left[A_\mu^{(1)}, A_\nu^{(2)}\right],$$

in which the superscripts (1), (2) and (3) are isospin indices and where the space-time subscripts are defined in the usual Minkowski notation

$$\partial_\mu := \frac{\partial}{\partial x_\mu}, \qquad x_\mu := (X, Y, Z, ict). \qquad (246)$$

The charge e in Eqs. (245) is *a quantized field quantity*, as discussed in detail later using the Cartesian X, Y and Z for the space indices. The presence of the Dirac constant $\hbar$ in the non-Abelian equations (245) is due to the usual quantum mechanical axiom (176) linking the momentum-energy four-vector p_μ to the wave four-vector κ_μ.

Therefore the non-Abelian definition of $F_{\mu\nu}$, Eqs. (245), automatically quantizes the electromagnetic field, producing the photon. In other words, the extension of electromagnetism from O(2) to O(3) produces quantization from fundamental geometry, an outcome which is an entirely natural consequence of the fact that space is three dimensional, and which is consistent with our contemporary appreciation of the vacuum itself as a geometrical concept with a non-trivial, non simply-connected, topology, based, not surprisingly, on O(3). It is this vacuum topology that allows the existence of the Aharonov-Bohm effect [16]. The equations (245) being a consequence of a rigorously geometrical theory of gauge fields [16] are consistent with gauge invariance of the first and second kind, and are the fundamental equations of the well known Yang-Mills theory of fields and particles.

3.1 GENERAL GEOMETRICAL THEORY OF GAUGE FIELDS

The essential difference between an O(2) and an O(3) theory of gauge fields is summarized through the fact that in O(2), rotation through an angle Λ_3 takes place in a plane, while in O(3) it takes place about an axis perpendicular to the plane. Rotations in a plane through an angle Λ_3 can be described by

$$\phi_1' = \phi_1 \cos \Lambda_3 + \phi_2 \sin \Lambda_3, \qquad \phi_2' = -\phi_1 \sin \Lambda_3 + \phi_2 \cos \Lambda_3, \qquad (247)$$

where ϕ_1 and ϕ_2 are field components [16]. However, the same rotation about an axis, 3, perpendicular to the plane requires the addition of

$$\phi_3' = \phi_3. \qquad (248)$$

The field therefore becomes a *vector* field ϕ with components (ϕ_1, ϕ_2, ϕ_3) *in three dimensional space*. In the U(1) (or O(2)) theory of electromagnetism, Eq. (248) is missing, so that the action is invariant under Eq. (247) only, i.e., invariant to a rotation in the (1,2) *plane* through an angle Λ_3. Rotations in two dimensions form the group O(2), which is also the group U(1). Thus gauge transformations of the first kind generate O(2) in this two dimensional world. Under a (1,2) plane rotation, a quantity Q is conserved, a quantity which

General Geometrical Theory of Gauge Fields 69

is identified as *unquantized* electric charge. Therefore in Abelian (U(1)) electrodynamics (AE) electric charge is not a quantized quantity. In *non-Abelian electrodynamics (NAE)*, which is indicated experimentally by $B^{(3)}$, it can be shown as follows *that charge is quantized* through the equation

$$e = \hbar\left(\frac{\kappa^2}{B^{(0)}}\right) = \hbar\left(\frac{\kappa}{A^{(0)}}\right), \qquad (249)$$

which supplements the usual quantum mechanical axioms,

$$En = \hbar\omega, \quad p = \hbar\kappa. \qquad (250)$$

Equation (249), derived later in this Section, is a third fundamental axiom of quantum mechanics, and shows that field-charge in NAE is subject to particle-wave dualism. In the same way that energy occurs in units of frequency, and linear momentum occurs in units of wave-number, e occurs in units of $\kappa/A^{(0)}$ or $\kappa^2/B^{(0)}$. The non-Abelian electromagnetic field is automatically quantized, and charge, e, becomes a property of the field itself, through $\kappa^2/B^{(0)}$. This type of charge quantization does not occur in O(2) electrodynamics, but is a direct consequence of the rigorous geometrical theory of non-Abelian gauge geometry applied to O(3). Charge quantization occurs in a three dimensional, but not in a two dimensional, theory of electromagnetism, illustrating the central importance of the field $B^{(3)}$ as observed in the inverse Faraday effect (Chap. 7 of Vol. 1) and other magnetic effects of light.

Note that Eq. (249) properly balances $\hat{C}$ symmetry (Chap. 2 of Vol. 1) and therefore conserves $\hat{C}$. In Eqs. (250), En and p are usually thought of as particulate (properties of matter), and ω and κ as undulatory (properties of waves). In AE, the electromagnetic wave is usually thought of as uncharged, in NAE, the field can act as its own source, and carries the *quantized* field charge defined by Eq. (249). In the static limit, $\kappa \to 0$; $B^{(0)} \to 0$ and e *remains finite*, becoming static, particulate, charge, the charge on the static electron. In this limit, there is no radiation (because there is no current, or moving charge) and so κ and $B^{(0)}$ are both zero. Equation (249) indicates that under $\hat{C}$ the sign of $B^{(0)}$ is reversed as well as that of e. In this view, there is no distinction between particulate and undulatory charge, an inference which is consistent with the view (Chap. 1) that action at a distance between charged particles (two electrons) takes place through the electromagnetic field or quantized photon. The latter is therefore *the agent of interaction at a distance* between the two electrons. Thus charge on the electron can be transmuted into the form of a field, an inference which is described in quantum mechanics by Eq. (249).

In analogy, gravitation in the theory of general relativity [16] is the agent of interaction at a distance between particulate point masses. The gravitational field itself carries energy, which is equivalent to mass and is *itself* a source of gravitation. In NAE, the electromagnetic field carries the quantized field charge defined by Eq. (249) and is its own source. An example of this is the by now familiar

$$\boldsymbol{B}^{(1)} \times \boldsymbol{B}^{(2)} = iB^{(0)}\boldsymbol{B}^{(3)*}, \quad \text{et cyclicum,} \qquad (251)$$

in which the non-Abelian cross product on the left hand side acts as a source for $\boldsymbol{B}^{(3)} = \boldsymbol{B}^{(3)*}$ *as the light travels in the vacuum*. Therefore, even in the absence of matter, the NAE field couples to itself, generating $\boldsymbol{B}^{(3)}$ in the vacuum. Analogously, in general relativity [16] the real divergence of the Einstein tensor $G_{\mu\nu}$ is non-zero, and the gravitational field *self generates*. Similarly, in considering the other components of Eq. (251), for example

$$\boldsymbol{B}^{(2)} \times \boldsymbol{B}^{(3)} = iB^{(0)}\boldsymbol{B}^{(1)*}, \qquad (252)$$

the cross product $\boldsymbol{B}^{(2)} \times \boldsymbol{B}^{(3)}$ is the source of $\boldsymbol{B}^{(1)*}$ in the vacuum, in the absence of matter. Finally, in the equation

$$\boldsymbol{B}^{(3)} \times \boldsymbol{B}^{(1)} = iB^{(0)}\boldsymbol{B}^{(2)*}, \qquad (253)$$

the cross product $\boldsymbol{B}^{(3)} \times \boldsymbol{B}^{(1)}$ becomes the source of $\boldsymbol{B}^{(2)*}$ in the vacuum. Light propagation through the vacuum therefore becomes understandable in terms of non-Abelian gauge geometry. There are direct analogies to this [16] in the curved space-time of general relativity. The cyclic relations (251), (252) and (253) are manifestations of the non-Abelian vertex diagram [16],

General Geometrical Theory of Gauge Fields 71

$$A \wedge A \wedge A \wedge A \wedge A \wedge A \wedge A \wedge A \wedge A \wedge A \wedge A \wedge A \wedge A \wedge A \wedge A \wedge A \quad (254)$$

i.e., the potential **A** of vacuum NAE *itself* emits gauge particles, transverse and longitudinal photons. In Eq. (254), A is now a vector with respect to *an isospin space* as well as a four-vector in space-time. Since $B^{(1)}$, $B^{(2)}$ and $B^{(3)}$ are known (Vol. 1) to be non-Abelian rotation generators in the group O(3), the isospin indices of NAE are (1), (2) and (3), and the conserved quantity of NAE becomes *isospin*, not the unquantized scalar charge of AE. Equations (251), (252) and (253) are therefore relations between vacuum magnetic field components in the isospin space defined by (1), (2) and (3). This space is also defined by the circular basis (221). The arbitrary isovector field [16] is defined in this basis by

$$\phi = \phi^{(1)} e^{(1)} + \phi^{(2)} e^{(2)} + \phi^{(3)} e^{(3)} = \phi_1 i + \phi_2 j + \phi_3 k. \quad (255)$$

In the general theory of gauge fields [16] an n component field ψ is subjected to the gauge transformation,

$$\psi(x_\mu) \to \psi'(x_\mu) = S(x_\mu)\psi(x_\mu), \quad (256)$$

where

$$S(x_\mu) = \exp(iM^a \Lambda^a(x_\mu)). \quad (257)$$

In these general gauge transformations, the isospin index a is still summed from one to three, but M^a are now $n \times n$ matrices representing group generators. For the O(3) group M^a are 3 X 3 matrices, the O(3) rotation generators (Vol. 1), obeying the Jacobi identity

$$[[M_i, M_j], M_k] + [[M_j, M_k], M_i] + [[M_k, M_i], M_j] = 0. \quad (258)$$

There emerges a deeply interesting parallel between non-Abelian electrodynamics and general relativity when we come to consider the transformation property [16]

Chapter 3. $B^{(3)}$ and Non-Abelian Gauge Geometry

$$\frac{\partial \psi'}{\partial x_\mu} = S\frac{\partial \psi}{\partial x_\mu} + \frac{\partial S}{\partial x_\mu}\psi. \qquad (259)$$

This is not a covariant transformation because ψ' is a function of isospin space as well as of the n dimensional space occupied by its n components. Therefore the infinitesimal $d\psi$ involves the variation of ψ with respect to *both* spaces, and is covariant if and only if the isospace axes are fixed, so that their infinitesimal variation is zero. The vector that results from this procedure, known as *parallel transport in isospace* [16] is denoted by $\psi + \delta\psi$. The latter is measured with respect to the local *iso*-coordinate system at $x_\mu + dx_\mu$ and is parallel to ψ measured in the local iso-coordinate system at x_μ. Therefore $\delta\psi$ is not zero because the local iso-coordinate systems are *different* at x_μ and $x_\mu + dx_\mu$. If the isocoordinate system is different, so is the vector itself, and $\psi \neq \psi + \delta\psi$. The general theory of gauge transformation geometry then proceeds [16] by assuming that

$$\delta\psi = i\frac{g}{\hbar}M^a A_\mu^a dx_\mu \psi, \qquad (260)$$

where g is a number, which in AE is the charge, e. The term A_μ^a describes to what extent the axes *in isospace* differ from point to point. The *true derivative* of ψ is now defined as

$$D\psi = (\psi + d\psi) - (\psi + \delta\psi) = d\psi - i\frac{g}{\hbar}M^a A_\mu^a dx_\mu \psi \qquad (261)$$

in Minkowski notation; the covariant derivative of the n dimensional field ψ transforming under a group whose generators are represented by the matrices M^a appropriate to the representation of ψ. Thus in S.I. units and Minkowski notation,

$$\frac{D\psi}{\partial x_\mu} := D_\mu\psi = \left(\frac{\partial}{\partial x_\mu} - i\frac{g}{\hbar}M^a A_\mu^a\right)\psi, \qquad (262)$$

is a correctly covariant derivative.

In Abelian (O(2)) electrodynamics, $M = -1$ and $g = e$, so that the isospin space is a scalar. In this case

General Geometrical Theory of Gauge Fields

$$D_\mu := \frac{\partial}{\partial x_\mu} + i\frac{e}{\hbar}A_\mu, \quad (263)$$

which is the (Cartesian) covariant derivative of AE. In NAE

$$M^1 = \begin{pmatrix} 0 & 0 & 0 \\ 0 & 0 & -i \\ 0 & i & 0 \end{pmatrix}, \quad M^2 = \begin{pmatrix} 0 & 0 & i \\ 0 & 0 & 0 \\ -i & 0 & 0 \end{pmatrix}, \quad M^3 = \begin{pmatrix} 0 & -i & 0 \\ i & 0 & 0 \\ 0 & 0 & 0 \end{pmatrix}, \quad (264)$$

in the O(3) group, and can be identified with the Cartesian components of the O(3) rotation generators used in Vol. 1,

$$J_X := M^1, \quad J_Y := M^2, \quad J_Z := M^3. \quad (265)$$

The elements of these matrices can be defined by

$$(M^a)_{mn} = -i\epsilon_{amn}. \quad (266)$$

It may be shown [16] that a vector rotated in isospin space produces the commutator $[D_\mu, D_\nu]$, through which may be defined *the non-Abelian field four-tensor* $G_{\mu\nu}$,

$$G_{\mu\nu} = \frac{\partial}{\partial x_\mu}\left(J^a A_\nu^a\right) - \frac{\partial}{\partial x_\nu}\left(J^a A_\mu^a\right) - i\frac{g}{\hbar}\left[J^a A_\mu^a, J^a A_\nu^a\right], \quad (267)$$

whose component form in the group O(3) is,

$$G_{\mu\nu}^{(3)*} = \frac{\partial A_\nu^{(3)*}}{\partial x_\mu} - \frac{\partial A_\mu^{(3)*}}{\partial x_\nu} - i\frac{g}{\hbar}\left[A_\mu^{(1)}, A_\nu^{(2)}\right], \quad (268a)$$

$$G_{\mu\nu}^{(2)*} = \frac{\partial A_\nu^{(2)*}}{\partial x_\mu} - \frac{\partial A_\mu^{(2)*}}{\partial x_\nu} - i\frac{g}{\hbar}\left[A_\mu^{(3)}, A_\nu^{(1)}\right], \quad (268b)$$

$$G_{\mu\nu}^{(1)*} = \frac{\partial A_\nu^{(1)*}}{\partial x_\mu} - \frac{\partial A_\mu^{(1)*}}{\partial x_\nu} - i\frac{g}{\hbar}\left[A_\mu^{(2)}, A_\nu^{(3)}\right]. \quad (268c)$$

These are equations (245) with the identity $e = g$; and become the familiar O(2) definition of $F_{\mu\nu}$ as the four-curl of A_μ in the Abelian O(2) group. The difference between Abelian and non-Abelian electrodynamics is embodied therefore in the non-zero commutators $[A_\mu^{(1)}, A_\nu^{(2)}]$, $[A_\mu^{(3)}, A_\nu^{(1)}]$, and $[A_\mu^{(2)}, A_\nu^{(3)}]$. In

74   Chapter 3. $B^{(3)}$ and Non-Abelian Gauge Geometry

O(2) electrodynamics, these commutators are *asserted* to be zero, whereas the cyclic relations (169), (172) and (189) show this assertion to be incorrect. In O(3) electrodynamics, the commutators are correctly taken to be non-zero.

3.1.1 THE QUANTIZATION OF CHARGE

The components of $G_{\mu\nu}$ must be electric and magnetic fields. The Z component of $B^{(3)}$ is now well defined through Eq. (268a), which reduces to

$$B_Z^{(3)} = G_{XY}^{(3)} = -G_{YX}^{(3)} = -i\frac{e}{\hbar}\left[A_X^{(1)}, A_Y^{(2)}\right]$$

(269)

$$= -i\frac{e}{\hbar}\left(A_X^{(1)}A_Y^{(2)} - A_Y^{(1)}A_X^{(2)}\right) = -i\frac{e}{\hbar}(\mathbf{A}^{(1)} \times \mathbf{A}^{(2)})_Z.$$

However, we know from Vol. 1 that

$$(\mathbf{A}^{(1)} \times \mathbf{A}^{(2)})_Z = i\frac{B^{(0)}}{\kappa^2}B_Z^{(3)},$$

(270)

which is a direct consequence of using the plane waves

$$\mathbf{A}^{(1)} = \mathbf{A}^{(2)*} = \frac{B^{(0)}}{\sqrt{2}\kappa}(i\mathbf{i} + \mathbf{j})e^{i\phi}.$$

(271)

Comparing Eqs. (269) and (270) gives the result

$$e = \hbar\frac{\kappa^2}{B^{(0)}},$$

(272)

which is the equation of *charge quantization* referred to earlier in section (3.1). Recall that this equation has been derived from the O(3) electrodynamics group with isospin indices (1), (2) and (3).

The alternative form of Eq. (272),

$$|\mathbf{p}| = eA^{(0)} = \hbar\kappa,$$

(273)

clarifies the nature of the quantized field charge,

$$e = \hbar\left(\frac{\kappa}{A^{(0)}}\right),$$

(274)

General Geometrical Theory of Gauge Fields 75

because $\hbar\kappa$ (and therefore $eA^{(0)}$) is the linear momentum magnitude of the free photon. In AE it is not customary to express the *free* photon momentum as $eA^{(0)}$ because this term is usually associated with the presence of charged matter (e.g. an electron as in Eq. (157) of Chap. 1). The free photon is on the other hand the quantized unit of light energy *in free space*. Equation (273) of NAE demonstrates that the linear momentum of the free photon *itself* is $eA^{(0)}$, where e is the field charge, an undulatory equivalent of particulate charge. This concept does not occur in AE.

3.1.2 ELECTRIC AND MAGNETIC FIELDS IN $G_{\mu\nu}$

Let us recall that the field $\boldsymbol{B}^{(3)}$ is implied by the cross product $\boldsymbol{B}^{(1)} \times \boldsymbol{B}^{(2)}$ of fields $\boldsymbol{B}^{(1)}$ and $\boldsymbol{B}^{(2)}$ *which also occur in AE*. Therefore $\boldsymbol{B}^{(1)}$ and $\boldsymbol{B}^{(2)}$ must take the same analytical form in AE as in NAE, and the four-tensor $G_{\mu\nu}$ must produce this result self consistently. In other words, $G_{\mu\nu}$ must contain the plane waves $\boldsymbol{B}^{(1)}$ and $\boldsymbol{B}^{(2)}$ together with $\boldsymbol{B}^{(3)}$, so that all three components are produced self-consistently from the same potential. The insights of this chapter have made it clear that this is not possible self-consistently within the structure of an O(2) theory, because the fundamental gauge geometry of O(2) does not allow non-zero commutators of the type appearing in $G_{\mu\nu}$ of Eq. (268).

In this section, Eq. (268) is developed to demonstrate its structure in detail (See Appendix D). The overall conclusions of this chapter thereby emerge as follows:

(1) In NAE, if the Abelian form of $\boldsymbol{A}^{(1)}$ and $\boldsymbol{A}^{(2)}$ is *assumed*, the field tensor $G_{\mu\nu}$ produces the Abelian $\boldsymbol{B}^{(1)}$, $\boldsymbol{B}^{(2)}$, $\boldsymbol{E}^{(1)}$ and $\boldsymbol{E}^{(2)}$ and self-consistently the intrinsically non-Abelian $\boldsymbol{B}^{(3)}$ and $-i\boldsymbol{E}^{(3)}/c$.

(2) In AE, the presence of $\boldsymbol{B}^{(3)}$ is *indicated* by $\boldsymbol{B}^{(1)}$ and $\boldsymbol{B}^{(2)}$ (as argued in Vol.1 and previous chapters of this volume), but the Abelian field tensor $F_{\mu\nu}$ does *not* produce $\boldsymbol{B}^{(3)}$ and $-i\boldsymbol{E}^{(3)}/c$ *self-consistently* from the Abelian $\boldsymbol{A}^{(1)}$ and $\boldsymbol{A}^{(2)}$. This is clear from the result

$$F_{12}^{(3)} = -F_{21}^{(3)} = \frac{\partial A_2^{(3)}}{\partial x_1} - \frac{\partial A_1^{(3)}}{\partial x_2} = 0. \quad (275)$$

In order to account for $\boldsymbol{B}^{(3)}$ we have had to *construct*

Chapter 3. $B^{(3)}$ and Non-Abelian Gauge Geometry

potential models, exemplified by that of Section 10.3 of Vol. 1.

(3) In NAE, the concept of wave-particle duality emerges for charge *itself* and the latter is quantized through Eq. (274). In essence, this means that in O(3) electrodynamics the photon is the agent of action at a distance between two electrons. The charge on the electron appears in the *free* photon momentum, which is defined as $eA^{(0)} = \hbar\kappa$. In O(2) electrodynamics, $\hbar\kappa$ is not conventionally identified with $eA^{(0)}$ for the free photon. Thus in NAE, the charge on the electron propagates as a wave-vector through free space.

In vector form, the Eqs. (268) can be written as

$$(G^{(1)*})_{\mu\nu} = (F^{(1)*})_{\mu\nu} - i\frac{e}{\hbar}(A^{(2)} \times A^{(3)})_{\mu\nu}, \qquad (276a)$$

$$(G^{(2)*})_{\mu\nu} = (F^{(2)*})_{\mu\nu} - i\frac{e}{\hbar}(A^{(3)} \times A^{(1)})_{\mu\nu}, \qquad (276b)$$

$$(G^{(3)*})_{\mu\nu} = (F^{(3)*})_{\mu\nu} - i\frac{e}{\hbar}(A^{(1)} \times A^{(2)})_{\mu\nu}. \qquad (276c)$$

These relations are true for each combination of μ and ν that defines an electric and magnetic field in $F_{\mu\nu}$. For example, $\mu\nu = 12, 21$, etc. define magnetic field components; $\mu\nu = 14, 41$, etc. define electric field components. Diagonal components vanish ($\mu\nu = 00, \ldots, 44$). Using the results (Eqs. (24) of Vol. 1),

$$A^{(1)} \times A^{(2)} = iA^{(0)}A^{(3)*} = i\frac{B^{(0)}}{\kappa^2}B^{(3)*},$$
$$\qquad (277)$$
$$A^{(2)} \times A^{(3)} = iA^{(0)}A^{(1)*}, \qquad A^{(3)} \times A^{(1)} = iA^{(0)}A^{(2)*},$$

and dropping the $\mu\nu$ subscripts in Eq. (276) for clarity of notation,

$$G^{(1)} = F^{(1)} - i\frac{e}{\hbar}A^{(0)}(iA^{(1)}),$$

$$G^{(2)} = F^{(2)} - i\frac{e}{\hbar}A^{(0)}(iA^{(2)}), \qquad (278)$$

$$G^{(3)} = F^{(3)} + \left(\frac{eB^{(0)}}{\hbar\kappa^2}\right)B^{(3)}.$$

Furthermore,

$$iA^{(1)} = \frac{iB^{(0)}}{\sqrt{2}\kappa}(i\mathbf{1}+\mathbf{J})e^{i\phi} = -\frac{E^{(1)}}{\omega},$$

$$iA^{(2)} = \frac{iB^{(0)}}{\sqrt{2}\kappa}(-i\mathbf{1}+\mathbf{J})e^{-i\phi} = \frac{E^{(2)}}{\omega},$$

(279)

and therefore Eq. (278) reduces to

$$G^{(1)} - F^{(1)} = \frac{iE^{(1)}}{c}, \quad G^{(2)} - F^{(2)} = -\frac{iE^{(2)}}{c},$$

$$G^{(3)} - F^{(3)} = B^{(3)},$$

(280)

a result which has been derived with the charge quantization conditions

$$e = \hbar\left(\frac{\kappa}{A^{(0)}}\right) = \hbar\left(\frac{\kappa^2}{B^{(0)}}\right).$$

(281)

Since O(3) electrodynamics is a theory of special relativity, the duality transformation can be applied to Eqs. (280), giving

$$G_D^{(1)} - F_D^{(1)} = B^{(1)}, \quad G_D^{(2)} - F_D^{(2)} = B^{(2)},$$

$$G_D^{(3)} - F_D^{(3)} = \frac{-iE^{(3)}}{c}.$$

(282)

Equations (280) to (282) show that in O(3) electrodynamics, the field $B^{(3)}$ and its dual $-iE^{(3)}/c$ emerge from the field four-tensor $G_{\mu\nu}$ in addition to $B^{(1)}$, $B^{(2)}$, $E^{(1)}$, and $E^{(2)}$, which retain their Abelian form — transverse plane waves. The isospin indices of O(3) electrodynamics are (1), (2) and (3) of the circular basis.

These conclusions represent a major development of our contemporary appreciation of electrodynamics. For example, the Maxwell equations *themselves* are generalized, as described in Chap. 4.

Chapter 4. The O(3) Maxwell Equations in the Vacuum

In this chapter the development of non-Abelian electrodynamics continues with an account of the O(3) Maxwell equations in the vacuum. It is shown that the charge quantization condition

$$e = \hbar \left(\frac{\kappa}{A^{(0)}} \right), \tag{283}$$

reduces the O(3) to the familiar O(2) vacuum Maxwell equations for the transverse fields $B^{(1)}$, $B^{(2)}$, $E^{(1)}$ and $E^{(2)}$. In the O(3) structure, however, there occur self-consistently Maxwell equations describing the vacuum spin fields $B^{(3)}$ and $-iE^{(3)}/c$. Electrodynamics is therefore the O(3), and not the O(2), sector of unified field theory. The traditional methods of developing the Maxwell equations are followed, i.e., they are divided into the vacuum inhomogeneous and vacuum homogeneous equations. It is important to note that throughout this chapter we shall be dealing with the Maxwell equations for free electromagnetism, i.e., do not discuss the interaction of free electromagnetism with matter. As seen in Chap. 3, O(3) electrodynamics implies that in the quantum interpretation, the linear momentum of the free photon can be described in terms of $eA^{(0)}$, where e is defined by Eq. (283). Unlike O(2) electrodynamics, the presence of this charge does *not* mean that there is an electron present, because in O(3) electrodynamics the field is its own source. This is precisely analogous with the well known Yang-Mills formulation as described for example by Ryder [16].

4.1 THE O(3) INHOMOGENEOUS MAXWELL EQUATIONS IN THE VACUUM

The O(2) inhomogeneous Maxwell equations (IME) are described in numerous textbooks. In Minkowski notation and S.I. units they are

Chapter 4. The O(3) Maxwell Equations

$$\frac{\partial F_{\mu\nu}}{\partial x_\nu} = 0, \qquad (284)$$

where $F_{\mu\nu}$ is the electromagnetic four-tensor, an antisymmetric tensor (Chap. 1) whose components are magnetic and electric fields. In three dimensional notation, Eqs. (284) are

$$\mathbf{\nabla}\cdot\mathbf{E} = 0, \qquad \mathbf{\nabla}\times\mathbf{B} = \frac{1}{c^2}\frac{\partial \mathbf{E}}{\partial t}, \qquad (285)$$

equations which show that there is no matter present, so that the source of free electromagnetism is infinitely distant, as in the traditional O(2) interpretation [54]. The IME are conventionally asserted to apply only to the transverse plane waves $\mathbf{B}^{(1)}$, $\mathbf{B}^{(2)}$, $\mathbf{E}^{(1)}$ and $\mathbf{E}^{(2)}$. Some texts mention that they also apply to phase free magnetic or electric fields, but these are discarded as irrelevant to plane *waves*. The development in Vol. 1 and Chaps. 1 to 3 has made this view obsolete, because the presence of vacuum plane waves implies the presence of the spin fields $\mathbf{B}^{(3)}$ and $-i\mathbf{E}^{(3)}/c$. The former is physical and produces a number of physical effects when free electromagnetism meets matter. Chapter 3 of this volume has shown that in O(3) electrodynamics, the free electromagnetic field can be thought of as carrying its own source, charge becomes quantized through Eq. (283). In the traditional O(2) development, the source of free electromagnetism must always be infinitely distant, so that the electromagnetism has taken an infinite time to reach the observer from the source at the speed of light. This makes O(2) electrodynamics inherently unsatisfactory in nature, because the radius of the known universe is thought to be finite, and no source can be infinitely distant from the earthbound observer. In the O(3) development, on the other hand, this difficulty is surmounted through Eq. (283), and furthermore, the philosophical basis for free electromagnetism becomes similar to that of free gravitation, as described briefly in Chap. 3. Thus, O(3) electrodynamics provides a natural bridge between electromagnetism and gravitation, a bridge that might close the gap between unified field theory and general relativity.

The O(3) counterpart of Eqs. (284) can be constructed by replacing the O(2) operator $\partial/\partial x_\mu$ by the O(3) operator D_μ, as defined in the *circular* basis in Appendix B:

The O(3) Inhomogeneous Maxwell Equations

$$D_\mu = \frac{\partial}{\partial x_\mu} + \frac{e}{\hbar}A_\mu^{(0)}. \tag{286}$$

The $F_{\mu\nu}$ tensor of O(2) electrodynamics must be replaced by the $G_{\mu\nu}$ tensor as described in Chap. 3. The $G_{\mu\nu}$ tensor is also a three component vector in isospin space, which is also the configuration space expressed in the circular basis (1), (2) and (3). As described in Chap. 3, this provides a self-consistent potential model for O(3) electrodynamics, a model which recognizes that the conjugate products $A^{(1)} \times A^{(2)}$; $B^{(1)} \times B^{(2)}$; and $E^{(1)} \times E^{(2)}$ are non-zero. As described in Chap. 3, O(2) electrodynamics self-indicates that it is incomplete and internally inconsistent, because these products can be expressed in terms of $B^{(3)}$ and $-iE^{(3)}/c$, fields which exist in the vacuum, and which add a third physical dimension to a planar (O(2) or U(1)) theory, making the latter obviously inconsistent with itself. Therefore, the O(3) IME equations are (see Appendix E)

$$D_\nu G_{\mu\nu} = 0, \tag{287}$$

which are equations for a vector in isospin space. The individual components (Appendix E) of this equation in the circular basis for the isospin (i.e., configuration) space are (Eq. B11),

$$D_\nu G_{\mu\nu}^{(1)} = 0, \quad D_\nu G_{\mu\nu}^{(2)} = 0, \quad D_\nu G_{\mu\nu}^{(3)} = 0. \tag{288}$$

From the outset, therefore, the O(3) theory is a theory in three physical (isospin) dimensions, and takes account of the existence of the spin fields $B^{(3)}$ and $-iE^{(3)}/c$ in the vacuum.

The charge quantization condition (Eq. (283) and Chap. 3) implies that in the vacuum

$$\frac{\partial}{\partial x_\mu} = \frac{e}{\hbar}A_\mu, \tag{289}$$

a condition (Appendix E) which reduces Eq. (287) to

$$\frac{\partial G_{\mu\nu}}{\partial x_\nu} = 0. \tag{290}$$

As described in Eqs. (280) of Chap. 3, the vacuum field tensor $G_{\mu\nu}$ can be expressed (Appendix D) as follows:

82   Chapter 4. The O(3) Maxwell Equations

$$G_{\mu\nu} = F_{\mu\nu} + F_{\mu\nu}, \qquad G_{\mu\nu}^{(D)} = F_{\mu\nu}^{(D)} + F_{\mu\nu}^{(D)}, \qquad (291)$$

where $F_{\mu\nu}^{(D)}$ is the dual tensor of $F_{\mu\nu}$ in special relativity. Equation (291) implies that the $\mu\nu$'th element of $G_{\mu\nu}$ is the corresponding element of $F_{\mu\nu}$ plus the identical element of $F_{\mu\nu}$. Since $F_{\mu\nu}$ and its dual $F_{\mu\nu}^{(D)}$ each obey the IME in the vacuum, we obtain from Eq. (287) and the charge quantization condition (283) the results

$$\frac{\partial F_{\mu\nu}}{\partial x_\nu} = 0, \qquad \frac{\partial F_{\mu\nu}^{(D)}}{\partial x_\nu} = 0, \qquad (292)$$

which contain the ordinary O(2) IME for the transverse, plane wave fields $B^{(1)}$, $B^{(2)}$, $E^{(1)}$ and $E^{(2)}$. In addition, Eqs. (292) contain the required IME equations for the spin fields $B^{(3)}$ and $-iE^{(3)}/c$.

In the O(2) theory of electrodynamics, charge, e, is conserved, while in the O(3) theory, the conserved quantity is $\hbar$, which is an angular momentum, or isospin. Thus in O(3) theory, the conserved quantity is isospin. Charge, e, is expressed in units of $\hbar$ through the condition (283), in the same way that the energy and linear and angular momenta of the free photon are expressed through units of $\hbar$. In O(3) theory it is possible, formally, to express the IME as (Appendix E),

$$\frac{\partial G_{\nu\mu}}{\partial x_\nu} = -e\left(\frac{A_\nu^{(0)} G_{\nu\mu}}{\hbar}\right) := J_\mu^{(eff)}, \qquad (293)$$

where $J_\mu^{(eff)}$ is an effective vacuum current, which is the *self-source* or *auto-source* of the propagating electromagnetic field. Since $\hbar$ appears in the charge quantization condition, the O(3) theory of electromagnetism is quantized from the outset. This follows from the fact that charge is quantized in O(3) electrodynamics.

4.2 THE O(3) HOMOGENEOUS MAXWELL EQUATIONS IN THE VACUUM

In the conventional, two dimensional, approach to vacuum electrodynamics the homogeneous Maxwell equations in the vacuum are ($\partial_\rho := \partial/\partial x_\rho$ etc.)

The O(3) Homogeneous Maxwell Equations

$$\partial_\rho F_{\mu\nu} + \partial_\mu F_{\nu\rho} + \partial_\nu F_{\rho\mu} = 0, \tag{294}$$

which in vector notation and S.I. units become

$$\nabla \cdot \mathbf{B} = 0, \qquad \nabla \times \mathbf{E} = -\frac{\partial \mathbf{B}}{\partial t}. \tag{295}$$

In the O(3) theory, the vacuum homogeneous Maxwell equations are

$$D_\rho \mathbf{G}_{\mu\nu} + D_\mu \mathbf{G}_{\nu\rho} + D_\nu \mathbf{G}_{\rho\mu} = \mathbf{0}, \tag{296}$$

whose components in the circular representation are

$$\begin{aligned}
\partial_\rho G^{(1)}_{\mu\nu} + \partial_\mu G^{(1)}_{\nu\rho} + \partial_\nu G^{(1)}_{\rho\mu} &= 0, \\
\partial_\rho G^{(2)}_{\mu\nu} + \partial_\mu G^{(2)}_{\nu\rho} + \partial_\nu G^{(2)}_{\rho\mu} &= 0, \\
\partial_\rho G^{(3)}_{\mu\nu} + \partial_\mu G^{(3)}_{\nu\rho} + \partial_\nu G^{(3)}_{\rho\mu} &= 0.
\end{aligned} \tag{297}$$

These equations allow self-consistently for the existence of the spin fields $\mathbf{B}^{(3)}$ and $-i\mathbf{E}^{(3)}/c$, while the corresponding O(2) equations apply conventionally to plane waves only. Using the charge quantization condition (289) and the condition (291) for $G_{\mu\nu}$ produces

$$\partial_\rho F^{(i)}_{\mu\nu} + \partial_\mu F^{(i)}_{\nu\rho} + \partial_\nu F^{(i)}_{\rho\mu} = 0, \qquad i = 1, 2, 3. \tag{298}$$

The third of these equations can be expressed in vector notation as

$$\nabla \cdot \mathbf{B}^{(3)} = 0, \qquad i\nabla \times \mathbf{E}^{(3)} = -\frac{\partial \mathbf{B}^{(3)}}{\partial t} = 0, \tag{299}$$

and shows that there is no Faraday induction due to $-\partial \mathbf{B}^{(3)}/\partial t$ in the vacuum, because the real and physical $\mathbf{B}^{(3)}$ is always linked through the homogeneous Maxwell Eq. (298) to the imaginary and unphysical $-i\mathbf{E}^{(3)}/c$, its dual in vacuo. Thus, chopping a circularly polarized laser beam in the vacuum will not produce a measurable, i.e., physical, electric field. However, such an induction is observed in the inverse Faraday effect, where $\mathbf{B}^{(3)}$ produces a real, physical magnetization in a material sample [55]. This magnetization relies on a property tensor, in the simplest case the susceptibility of a single electron, as described in Chap. 12 of Vol. 1 through the relativistic Hamilton-Jacobi equation. If there is no electron present, that equation shows that there is no magnetization, no inverse Faraday effect, and therefore no electric field due to Faraday induction, and no current in the measuring induction coil [55]. Seen in another way, the field $-i\boldsymbol{E}^{(3)}/c$ is pure imaginary in the vacuum, has no real part, and is therefore unphysical. It is an electric field whose real part is zero, and its curl is always zero. Therefore the vacuum $-\partial \boldsymbol{B}^{(3)}/\partial t$ is also zero from Eq. (299), and the vacuum $\boldsymbol{B}^{(3)}$ is a constant. Analogously, the angular momentum of the free photon, $\hbar$, is a constant. Special relativity asserts that a longitudinal, non-zero electric field $i\boldsymbol{E}^{(3)}$ in the vacuum is invariant under Lorentz transformation (Vol. 1), so that its curl in the vacuum is zero. Special relativity also implies that a longitudinal axial vector such as $\boldsymbol{B}^{(3)}$ is relativistically invariant, and cannot change with time. This is another way of saying that there cannot be Faraday induction due to $\partial \boldsymbol{B}^{(3)}/\partial t$ in the vacuum.

The O(3) homogeneous Maxwell equations in the vacuum can be expressed formally (Appendix E) as

$$\partial_\rho G_{\mu\nu} + \partial_\mu G_{\nu\rho} + \partial_\nu G_{\rho\mu} = -\frac{e}{\hbar}\left(A_\rho^{(0)} G_{\mu\nu} + A_\mu^{(0)} G_{\nu\rho} + A_\nu^{(0)} G_{\rho\mu}\right), \qquad (300)$$

and therefore allow formally for the existence of a magnetic monopole. This conclusion is well known [16] in the context of 't Hooft Polyakov monopoles, which are derived by fixing one isospin axis in configuration space and modelling the potential A_μ.

4.3 THE DUALITY TRANSFORMATION AND THE O(3) MAXWELL EQUATIONS

We have seen that the O(3) Maxwell equations supplement the O(2) equivalents with the vacuum equations,

$$\nabla \times \boldsymbol{B}^{(3)} = \frac{1}{c^2}\frac{\partial \boldsymbol{E}^{(3)}}{\partial t} = 0, \qquad \nabla \times \boldsymbol{E}^{(3)} = -\frac{\partial \boldsymbol{B}^{(3)}}{\partial t} = 0, \qquad (301)$$

$$\nabla \cdot \boldsymbol{E}^{(3)} = 0, \qquad \nabla \cdot \boldsymbol{B}^{(3)} = 0.$$

In Vol. 1 and in previous chapters of this volume it has been asserted that $\boldsymbol{E}^{(3)}$ is pure imaginary if $\boldsymbol{B}^{(3)}$ is pure real. Thus, there can be no physical effect of $\boldsymbol{E}^{(3)}$, in contrast to

$B^{(3)}$, which is observable in many different ways. In classical electrodynamics, $B^{(3)}$ and $E^{(3)}$ from the O(3) Maxwell equations are purely irrotational and divergentless, and since $E^{(3)}$ is pure imaginary (and henceforth denoted $-iE^{(3)}$ to emphasize this property) there can be no Faraday induction *in free space* due to $\nabla \times B^{(3)}$, whose real part is always zero. (As described already, however, $B^{(3)}$ sets up a real, physical, and measurable magnetization in material matter, at its simplest one electron, and this real magnetization can be detected by an induction coil, as in the original series of experiments by van der Ziel *et al.* [55] which demonstrated the inverse Faraday effect.)

This section discusses the duality transformation of special relativity as applied to O(3) electrodynamics, beginning with the duality transformation applied to Eqs. (301). It is deduced that $-iE^{(3)}/c$ is dual to $B^{(3)}$ within O(3), non-Abelian, electrodynamics, meaning that if $B^{(3)}$ is real and physical, $-iE^{(3)}/c$ is imaginary and unphysical. This result reinforces the general symmetry arguments of Vol. 1, elementary arguments which show that a polar vector, such as a real electric field, cannot be obtained from the vector cross product of two other polar vectors or two other axial vectors. Thus, the conjugate product $E^{(1)} \times E^{(2)} = c^2 B^{(1)} \times B^{(2)}$ cannot produce a real electric field in vacuo, it always produces a real *axial* vector, a real and physical magnetic field, $B^{(3)}$. This property means that the group of electrodynamics is the non-Abelian O(3), not the Abelian and planar O(2), since within O(2), there is no allowance made for a physical field in an axis orthogonal to the plane of O(2). This deduction means that quantum electrodynamics (Chap. 6) must also be adjusted to account for $B^{(3)}$, and similarly for unified field theory (Chap. 5).

4.3.1 THE DUAL OF $B^{(3)}$ IN O(3) ELECTRODYNAMICS

The dual of $B^{(3)}$ is not zero, despite the fact that symmetry forbids a real electric field in the (3) state. It is shown in this section that it is a pure imaginary, vacuum electric field $-iE^{(3)}/c$ which is unphysical according to the rule that real fields are physical and vice versa. The only physical effect of fields in state (3) occurs through $B^{(3)}$, as shown already through the fundamental classical and quantum equations governing the interaction of the electromagnetic field with one electron, the simplest representation of matter. There appears to be no experimental evidence for a real $\mathbf{E}^{(3)}$, but there is plentiful evidence, discussed already in these volumes, for the existence of $\mathbf{B}^{(3)}$.

The duality transformation of special relativity when applied to $\mathbf{B}^{(3)}$ asserts that the Maxwell equations (301) of O(3) gauge theory are invariant under the transformations

$$\mathbf{B}^{(3)} \to \frac{-i\mathbf{E}^{(3)}}{c}, \qquad \mathbf{E}^{(3)} \to ic\mathbf{B}^{(3)}, \qquad (302)$$

i.e., the same equations are obtained if $\mathbf{B}^{(3)}$ is replaced by $-i\mathbf{E}^{(3)}/c$ and if $\mathbf{E}^{(3)}$ is replaced by $ic\mathbf{B}^{(3)}$ (in S.I. units). The duality transformation (302) is a fundamental property of tensors in Minkowski space-time, and means that if $\mathbf{B}^{(3)} := B^{(0)}\mathbf{k}$, i.e., is defined as real, then $B^{(0)} \to -iE^{(0)}/c$ gives the imaginary $\mathbf{E}^{(3)} = iE^{(0)}\mathbf{k}$; and if $\mathbf{E}^{(3)} := iE^{(0)}\mathbf{k}$, then $E^{(0)} \to icB^{(0)}$ gives the real $\mathbf{B}^{(3)} = B^{(0)}\mathbf{k}$. This kind of transformation is discussed in Appendix C of Vol. 1. The O(3) Maxwell equations (301) are therefore invariant under the duality transformation if and only if $\mathbf{B}^{(3)}$ is dual to $-i\mathbf{E}^{(3)}/c$ and $\mathbf{E}^{(3)}$ is dual to $ic\mathbf{B}^{(3)}$. In O(3) vacuum electrodynamics, however, the cyclic relations (251) between three physical fields ensure that $\mathbf{B}^{(3)}$ is pure real, so is dual to the pure imaginary $-i\mathbf{E}^{(3)}/c$.

This result is reinforced when we consider the tensorial form of the inhomogeneous part of Eq. (301)

$$\frac{\partial F^{(3)}_{\mu\nu}}{\partial x_\mu} = 0, \qquad (303)$$

where

$$F^{(3)}_{\mu\nu} = \begin{bmatrix} 0 & cB^{(3)}_z & 0 & 0 \\ -cB^{(3)}_z & 0 & 0 & 0 \\ 0 & 0 & 0 & -iE^{(3)}_z \\ 0 & 0 & iE^{(3)}_z & 0 \end{bmatrix}, \qquad (304)$$

with $B^{(3)}_z = B^{(0)}$, $E^{(3)}_z = E^{(0)}$.

Written out in full, Eqs. (303) are

$$\frac{\partial F_{12}}{\partial x_2} = c\frac{\partial B^{(3)}_z}{\partial Y} = 0, \qquad (305a)$$

Duality Transformation and O(3) Maxwell Eqs.

$$\frac{\partial F_{21}}{\partial x_1} = -c\frac{\partial B_Z^{(3)}}{\partial X} = 0, \quad (305b)$$

$$\frac{\partial F_{34}}{\partial x_4} = -i\frac{\partial E_Z^{(3)}}{ic\partial t} = 0, \quad (305c)$$

$$\frac{\partial F_{43}}{\partial x_3} = i\frac{\partial E_Z^{(3)}}{\partial Z} = 0. \quad (305d)$$

Equation (305c) allows the electric field to be either pure real, i.e., $-\partial E_Z^{(3)}/\partial t = 0$ or imaginary, i.e., $-\partial i E_Z^{(3)}/\partial t = 0$, and the latter choice is taken because the O(3) cyclic relations (251) demand a real magnetic field as discussed already. This is the basis of our statement that $\boldsymbol{B}^{(3)}$ is dual to $-i\boldsymbol{E}^{(3)}/c$ in O(3) vacuum electrodynamics. The Lorentz invariant [56],

$$L^{(3)} = F_{\mu\nu}^{(3)} F_{\mu\nu}^{(3)} = 0, \quad (306)$$

therefore vanishes in the vacuum, but if and only if $\boldsymbol{B}^{(3)}$ is accompanied by $-i\boldsymbol{E}^{(3)}/c$. Similarly, the invariants from the accompanying plane waves represented by states (1) and (2) also vanish in the vacuum. Specifically,

$$L^{(1)} = F_{\mu\nu}^{(1)} F_{\mu\nu}^{(1)} = 0, \quad L^{(2)} = F_{\mu\nu}^{(2)} F_{\mu\nu}^{(2)} = 0, \quad (307)$$

where

$$F_{\mu\nu}^{(1)} = \begin{bmatrix} 0 & 0 & -cB_Y^{(1)} & -iE_X^{(1)} \\ 0 & 0 & cB_X^{(1)} & -iE_Y^{(1)} \\ cB_Y^{(1)} & -cB_X^{(1)} & 0 & 0 \\ iE_X^{(1)} & iE_Y^{(1)} & 0 & 0 \end{bmatrix}, \quad (308)$$

and similarly for $F_{\mu\nu}^{(2)}$. The Cartesian components in Eq. (308) are given by

$$\boldsymbol{B}^{(1)} := B_X^{(1)}\boldsymbol{i} + B_Y^{(1)}\boldsymbol{j}, \quad \boldsymbol{E}^{(1)} := E_X^{(1)}\boldsymbol{i} + E_Y^{(1)}\boldsymbol{j}, \quad (309)$$

and similarly for $B^{(2)}$ and $E^{(2)}$, where

$$B_X^{(1)} = \frac{iB^{(0)}}{\sqrt{2}} e^{i\phi}, \qquad B_Y^{(1)} = \frac{B^{(0)}}{\sqrt{2}} e^{i\phi},$$
$$E_X^{(1)} = \frac{E^{(0)}}{\sqrt{2}} e^{i\phi}, \qquad E_Y^{(1)} = -\frac{iE^{(0)}}{\sqrt{2}} e^{i\phi}. \qquad (310)$$

For the circular state (2), these Cartesian components are given by the complex conjugates of those in Eq. (110), i.e., by

$$B_X^{(2)} = -i\frac{B^{(0)}}{\sqrt{2}} e^{-i\phi}, \qquad B_Y^{(2)} = \frac{B^{(0)}}{\sqrt{2}} e^{-i\phi},$$
$$E_X^{(2)} = \frac{E^{(0)}}{\sqrt{2}} e^{-i\phi}, \qquad E_Y^{(2)} = i\frac{E^{(0)}}{\sqrt{2}} e^{-i\phi}. \qquad (311)$$

Thus, equations (309) represent complex Cartesian components of circular states (1) and (2) respectively. These components occur in vacuum O(2) electrodynamics, and are unchanged, as we have seen, in O(3) electrodynamics. However, Eq. (301) is self consistent if and only if the gauge group of electrodynamics is O(3). Clearly, the components in $F_{\mu\nu}^{(1)}$ and $F_{\mu\nu}^{(2)}$ are complex and oscillatory in general (i.e., contain real and imaginary parts and depend on the phase ϕ), whereas those in $F_{\mu\nu}^{(3)}$ are either pure real or pure imaginary, and are phase free, i.e., independent of ϕ.

The duality transformation applied to $F_{\mu\nu}^{(1)}$ or $F_{\mu\nu}^{(2)}$ also works in a slightly different way. For $F_{\mu\nu}^{(1)}$,

$$B^{(1)} \rightarrow -\frac{iE^{(1)}}{c}, \text{ and } E^{(1)} \rightarrow icB^{(1)}, \qquad (312)$$

is equivalent to

$$B^{(0)} \rightarrow -\frac{E^{(0)}}{c}, \text{ and } E^{(0)} \rightarrow -cB^{(0)}. \qquad (313)$$

For $F_{\mu\nu}^{(2)}$,

$$B^{(2)} \rightarrow -\frac{iE^{(2)}}{c}, \text{ and } E^{(2)} \rightarrow icB^{(2)}, \qquad (314)$$

is equivalent to

Duality Transformation and O(3) Maxwell Eqs.

$$B^{(0)} \to \frac{E^{(0)}}{c}, \text{ and } E^{(0)} \to cB^{(0)}. \tag{315}$$

On the other hand

$$\boldsymbol{B}^{(3)} \to -\frac{i\boldsymbol{E}^{(3)}}{c}, \text{ and } \boldsymbol{E}^{(3)} \to ic\boldsymbol{B}^{(3)}, \tag{316}$$

is equivalent to

$$B^{(0)} \to -\frac{iE^{(0)}}{c}, \text{ and } E^{(0)} \to icB^{(0)}. \tag{317}$$

These features mean that $\boldsymbol{B}^{(1)}$ and $\boldsymbol{E}^{(1)}$ are complex (i.e., have both real and imaginary parts) and so are $\boldsymbol{B}^{(2)}$ and $\boldsymbol{E}^{(2)}$, whereas $\boldsymbol{B}^{(3)}$ is pure real and $-i\boldsymbol{E}^{(3)}$ pure imaginary. The Cartesian components in the circular state (3) are

$$\boldsymbol{B}^{(3)} = B_Z^{(1)}\boldsymbol{k}, \quad -i\boldsymbol{E}^{(3)} = -iE_Z^{(3)}\boldsymbol{k}, \tag{318}$$

with

$$B_Z^{(1)} = B^{(0)}, \quad E_Z^{(3)} = E^{(0)}. \tag{319}$$

In O(2) electrodynamics, the components (318) are unconsidered, in O(3) electrodynamics, they occur self-consistently with the Cartesian components of the circular states (1) and (2). The development of classical O(3) electrodynamics has repercussions in QED and unified field theory which are described later in this volume; the occurrence of the physical $\boldsymbol{B}^{(3)}$ field is self-indicated in O(2) through the conjugate product $\boldsymbol{B}^{(1)} \times \boldsymbol{B}^{(2)} = iB^{(0)}\boldsymbol{B}^{(3)*}$, and shows up in the classical, relativistic, Hamilton-Jacobi equation of one electron in the electromagnetic field (Vol. 1, Chap. 12).

4.4 RENORMALIZATION OF O(3) QED

The occurrence of $\boldsymbol{B}^{(3)}$ in classical electrodynamics in the vacuum means that the gauge group of electromagnetism is O(3) throughout field theory, specifically in QED and unified field theory. It is well known [16] that O(3) gauge theories are renormalizable in QED. Therefore, without giving details, it is inferred that the vacuum O(3) Maxwell equations of this chapter are renormalizable in QED, essentially because the power counting argument allows this to be so. The number of primitively divergent graphs is finite in O(3) gauge theories when spontaneous symmetry breaking (SSB) is absent, and the gauge, ghost and source fields can be rescaled, meaning that the generating function Γ is finite as $\epsilon \to 0$ and that gauge invariance is preserved at each order. The generating function in O(3) QED can always be made finite by the addition of counter-terms to obtain renormalized generating functionals. For example, the zero and one loop functionals can be made finite by the addition of counter-terms that obey the Slavnov-Taylor identities. In O(3) QED, the only type of interaction between matter and the gauge field is a vector interaction of the form $g_v J_\mu W_\mu$ where W_μ is the gauge field and J_μ the vector current of the Fermi matter field.

The vector current is conserved, leading to a Ward identity for the vertex function. In non-Abelian gauge theory, the Ward identity is generalized, and is essential to the proof of the renormalizability of the gauge theory. In unified field theory, however, renormalizability is assured by the fact that the total contribution of the triangle graphs is zero, so triangle anomalies cancel. This is a condition on the fermion content of the theory, which is satisfied if there exist quarks as well as leptons, and if the quarks carry an additional SU(3) color label. It is also well known that SSB does not affect the renormalizability of Abelian and non-Abelian gauge theories using the 't Hooft gauge, which introduces an effective potential. In incorporating $B^{(3)}$ into unified field theory, therefore, care must be taken to ensure that renormalizability is maintained, but since $B^{(3)}$ occurs in *classical* electrodynamics, it must be described consistently in QED (Chap. 6) and unified field theory (Chap. 5). Therefore massive vector particles, such as vector bosons, do not destroy renormalizability. The existence of the intermediate vector boson of the Weinberg Salam (WS) model actually depends on the SSB of a non-Abelian gauge theory, and it is therefore natural to enquire, as in Chap. 5, how O(3) electrodynamics affects this model. At first glance, WS is presumably extended from SU(2) ⊗ U(1) to SU(2) ⊗ SU(2) (the Lorentz group), or some other variant which would incorporate $B^{(3)}$ and preserve the ability of WS to reproduce the correct vector boson masses. The very name *vector boson* arises out of non-Abelian gauge theory.

Work of this kind is already available [16] in the theoretical anticipation of 't Hooft Polyakov (HP) magnetic monopoles. This is based on enlarging the gauge symmetry of electromagnetism [16] from O(2) to O(3), while simultaneously introducing SSB (Chap. 2). The theory can be developed with an O(3) symmetry group with a gauge field $F^a_{\mu\nu}$, where a is an isovector index of the type used in this chapter, i.e.,

$$F^a_{\mu\nu} = \partial_\mu A^a_\nu - \partial_\nu A^a_\mu + \frac{e}{\hbar}\epsilon_{abc}A^b_\mu A^c_\nu, \tag{320}$$

from which magnetic monopoles are obtained by introduction of a Higgs field and by modelling the vector potential. It is therefore natural to picture $\boldsymbol{B}^{(3)}$ as being produced by two magnetic monopoles of the HP variety in O(3) electrodynamics. However, there is a well known difficulty [16] with the HP monopole in that it cannot be obtained from the WS model of unified field theory, essentially because the conventional U(1) (or O(2)) sector is irregularly imbedded and non-compact. Therefore HP monopoles do not exist in the WS model, a situation which however, might be remedied if the electromagnetic sector is enlarged to O(3). The latter is a description of three dimensional space, as is SU(2) (Chap. 1), while O(2) is a description of a flat space. Magnetic monopoles might well be reinstated in a unified theory based on an O(3) electromagnetic sector. These considerations are left to Chap. 5.

It is also known that the gauge group SU(2) plays a role in unifying the concepts behind the Dirac and HP monopoles, and also in the derivation of instantons [16], solitons (Chap. 2) which are localized in space and time. (We note that the quantized version of $\boldsymbol{B}^{(3)}$, the photomagneton [1–10] is also localized in space and time.) The SU(2) gauge group allows instantons to exist, furthermore, in the absence of SSB, and the form of $F_{\mu\nu}$ in instanton theory is exactly the same as in Eq. (320) of this chapter, so $\boldsymbol{B}^{(3)}$ is well defined in instanton theory, being based in O(3) electrodynamics on the same equation for $F_{\mu\nu}$, provided that the isospin indices are the configuration space indices (1), (2) and (3) of the circular basis. The instanton solution to non-Abelian field equations represents a transition from one class of Yang Mills vacua to another. The Yang Mills vacuum is infinitely degenerate, consisting of an infinite number of homotopically non-equivalent vacua. Finally, it is known that non-Abelian gauge theories that occur in electroweak theory, QCD, SU(5) and possibly, general relativity are richly structured with many physical insights. The existence of $\boldsymbol{B}^{(3)}$ in classical electrodynamics is the key to unification of electrodynamics with other concepts based on these non-Abelian geometries.

4.5 ISOSPIN AND GAUGE SYMMETRY

In the rotation group O(3), isospin is a conserved quantity, a vector quantity, angular momentum. The charge quantization equation (283) relates the angular momentum magnitude $\hbar$ to charge, the conserved quantity of O(2) electrodynamics. This is achieved by identifying isospin space with configuration space (1), (2) and (3) in the circular basis. The third circular state (3) is the spin state associated with the experimentally observable field $\boldsymbol{B}^{(3)}$. The latter therefore shows that isospin symmetry is a local (or gauge) symmetry. In O(3) electrodynamics, type one gauge transformations are rotations in the configuration space (1), (2), (3); whereas in O(2) electrodynamics the space is flat, i.e., the only circular indices considered in O(2) are (1) and (2). The conjugate product $\boldsymbol{B}^{(1)} \times \boldsymbol{B}^{(2)}$ of O(2) theory indicates, however, that this flat space is not self-consistent, because the conjugate product produces a physical field, $\boldsymbol{B}^{(3)}$, which exists in an axis *orthogonal* to the flat O(2) plane, and cannot therefore be in the plane. It follows that self-consistency in classical electrodynamics can be achieved only with a gauge theory that is not O(2), and the simplest generalization is O(3). In this view therefore electromagnetism, color, weak isospin, and hypercharge, are all non-Abelian concepts. Isospin is generally understood to mean a conserved vector quantity and isospin space to mean an internal symmetry space such as the configuration space (1), (2) and (3) used in this chapter. The vector quantity $F_{\mu\nu}$ of O(3) electrodynamics carries isospin (I = 1) and by definition acts as a source for itself in the vacuum. The field $F_{\mu\nu}$ becomes a consequence of the existence of a particle with isospin, which in O(3) is identifiable as the photon. This is a direct consequence [16] of the fact that the symmetry group O(3) is non-Abelian. In O(3) electrodynamics therefore, the electromagnetic field itself may emit gauge particles and be self-propagating, in direct analogy (Appendix C) with the gravitational field. This analogy may ultimately allow the unification of electromagnetism with gravitation. Feynman rules for non-Abelian gauge fields are well defined in QED, and Faddeev-Popov ghosts [16] can be integrated out using the axial gauge. Ward identities can be satisfactorily generalized to the non-Abelian case in QED, so there is no fundamental objection to the development of $\boldsymbol{B}^{(3)}$ in QED, a development which will probably lead to a much richer electrodynamical structure than O(2), provided each stage of theoretical development is checked against the available data, particularly much needed

Isospin and Gauge Symmetry

data on magnetization by light and electromagnetic radiation.

Finally, non-Abelian theories allow charge quantization, because all isovector fields couple with the same strength to the gauge field through a coupling constant g. If this is identified with electric charge, as in this chapter, Eq. (283) identifies this as the field charge, e, and identifies photon momentum in O(3) as $eA^{(0)} = \hbar\kappa$. The existence of this equation is another direct consequence of the use of O(3) gauge geometry for electrodynamics, which in turn is a direct consequence of the existence of $B^{(3)}$ through the conjugate product. The emergence of $B^{(3)}$ from the relativistic equations of one electron in the classical electromagnetic field is conclusive evidence that O(2) electrodynamics is internally inconsistent. It is of the utmost contemporary importance to devise accurate experimental detection of the characteristic square root power density dependence of $B^{(3)}$ in the radio frequency magnetization of an electron plasma as described in simple one electron terms in Chap. 12 of Vol. 1. Since $B^{(1)} \times B^{(2)}$ of O(2) theory is observable in the same experiment, there is no reasonable doubt as to the existence of $B^{(3)}$ itself as an experimental observable. In other words there is no way in which $B^{(3)}$ can be zero, because the non-zero observable $B^{(1)} \times B^{(2)} = iB^{(0)}B^{(3)*}$. It follows immediately that O(2) electrodynamics is internally self-contradictory. This chapter has shown that O(3) electrodynamics incorporates the vacuum $B^{(3)}$ self consistently. In the next chapter we explore the consequences of an O(3) electromagnetic sector for WS unified field theory.

Chapter 5. $B^{(3)}$ in Unified Field Theory

In previous chapters we have argued that the U(1) sector of unified field theory must be replaced by an O(3) symmetry, and that this affects unified field theory. These assertions are based on the experimental observation of the electromagnetic field $B^{(3)}$ in magnetization by light. The enlargement of electromagnetic symmetry to O(3) must, moreover, be carried out in such a way that preserves the ability of unified field theory to reproduce experimental data in particle physics, as in the well known CERN experiment in 1983 [57] which verified the theory of Weinberg and Salam [16], henceforth referred to as *GWS* unified field theory. The product group of GWS is SU(2) ⊗ U(1), the SU(2) sector being non-Abelian. Masses of novel intermediate vector bosons are predicted by the theory with the use of spontaneous symmetry breaking (SSB), sketched out in Chap. 2. The gauge bosons, $W^{\pm}$, which mediate the weak interaction [16] in GWS are vector bosons, which arise from non-Abelian symmetry. They are introduced as three gauge potentials, W_μ^i, which carry a Cartesian isospin index i and which appear in the covariant derivative of an isospinor L. The latter is defined as the isospinor

$$L := \begin{pmatrix} \mathbf{v}_e \\ e_L \end{pmatrix}, \tag{321}$$

where $\mathbf{v}_e$ is a left handed electron neutrino and e_L a left handed electron. These particles have the same space-time properties and can therefore be used in an isospinor with the same parity. The doublet defined by the isospinor (321) has the non-Abelian charge $I_W = 1/2$ where I_W is weak isospin. Under SU(2) gauge transformation, the isospinor transforms as

$$L \to e^{-(i/2)\boldsymbol{\tau}\cdot\boldsymbol{\alpha}} L := \boldsymbol{S} L, \tag{322}$$

where $\boldsymbol{S}$ is a 2 x 2 matrix and $\boldsymbol{\tau}/2$ are the Pauli matrices of

Chap. 1. Since there is no right handed neutrino (from parity violation [58] experiments on beta decay), the right handed electron does not form an isospinor with a neutrino, and is represented in GWS by an isosinglet,

$$R = e_R. \tag{323}$$

This is represented in the non-Abelian charge state $I_W = 0$. This should not be confused with the non-Abelian coupling constant g, which is identified in magnetic monopole theory with the electronic charge e. In this representation, the neutrino is labelled $I_W^3 = 1/2$, and the left handed electron by $I_W^3 = -1/2$. The relation between electric charge, Q, and these non-Abelian conserved quantities is given by [16]

$$L := Q = I_W^3 - \frac{1}{2}, \qquad R := Q = I_W^3 - 1, \tag{324}$$

so that the neutrino has zero electric charge in this non-Abelian representation. Under gauge transformation, the pertinent Cartesian covariant derivatives are

$$D_\mu L = \partial_\mu L - \frac{i}{2} g \boldsymbol{\tau} \cdot \boldsymbol{W}_\mu L, \qquad D_\mu L = \partial_\mu L + \frac{i}{2} g' X_\mu L,$$

$$D_\mu R = \partial_\mu R + i g' X_\mu R, \tag{325}$$

where g and g' are coupling constants [16]. These arise in conventional GWS because e_R can be subjected to a U(1) transformation, while the isospin doublet (321) undergoes an SU(2) transformation. The X_μ potentials in Eqs. (325) are therefore consequences of a U(1) gauge transformation within the structure of GWS theory. If this were to be generalized to an SU(2) (or O(3)) transformation, the overall symmetry of GWS would become that of the Lorentz group SU(2) ⊗ SU(2). This U(1) gauge symmetry in conventional GWS leads to a conserved weak hypercharge,

$$Q = I_W^3 + \frac{Y_W}{2}, \tag{326}$$

as first suggested by Weinberg [59]. The gauge field corresponding to this U(1) symmetry is therefore not the photon field, but X_μ (and $\boldsymbol{W}_\mu$) become parts of the electromagnetic four-potential

$$A_\mu = \frac{g'W_\mu^3 + gX_\mu}{(g^2 + g'^2)^{\frac{1}{2}}} := W_\mu^3 \sin\theta_W + X_\mu \cos\theta_W. \tag{327}$$

The U(1) transformation which gives rise to X_μ is

$$e_R \to e^{i\beta} e_R, \tag{328}$$

and is therefore a transformation on the right handed electron e_R. Clearly, the above is the briefest of sketches of GWS theory, but is enough to show that its overall structure is based on gauging in SU(2) and U(1). It is clear, however, that GWS theory gives the results

$$\begin{aligned} \mathbf{A}^{(1)} &= \mathbf{W}_3^{(1)} \sin\theta_W + \mathbf{X}^{(1)} \cos\theta_W, \\ \mathbf{A}^{(2)} &= \mathbf{W}_3^{(2)} \sin\theta_W + \mathbf{X}^{(2)} \cos\theta_W, \end{aligned} \tag{329}$$

where θ_W is the Weinberg angle defined by

$$\theta_W = \cos^{-1}\frac{g}{(g^2 + g'^2)^{\frac{1}{2}}}. \tag{330}$$

Therefore the experimentally observable field $\mathbf{B}^{(3)}$ is given by

$$\begin{aligned} \mathbf{B}^{(3)} &= -i\frac{\kappa^2}{B^{(0)}} \mathbf{A}^{(1)} \times \mathbf{A}^{(2)} = -i\frac{\kappa^2}{B^{(0)}} \\ &\quad \times \left(\mathbf{W}_3^{(1)} \sin\theta_W + \mathbf{X}^{(1)} \cos\theta_W\right) \times \left(\mathbf{W}_3^{(2)} \sin\theta_W + \mathbf{X}^{(2)} \cos\theta_W\right), \end{aligned} \tag{331}$$

which involves cross products such as $\mathbf{W}_3^{(1)} \times \mathbf{W}_3^{(2)}$, $\mathbf{X}^{(1)} \times \mathbf{X}^{(2)}$, and cross terms. In conventional GWS, however, W_μ^3 is fixed in isospin axis 3 and A_μ and X_μ are isospin scalars. Isospin space is *not* identified with configuration space in conventional GWS and the Higgs field which is used in the theory [16] is fixed in isospin axis three.

With the introduction of isospin indices (1), (2) and (3) as in Chaps. 3 and 4 for $\mathbf{A}_\mu^{(1)}$, however, the potential functions W_μ^3 and X_μ of conventional GWS must also take on the same isospin indices, which become indices of the circular basis of physical, three dimensional space. Although the Weinberg angle is fixed experimentally [16] as $\theta_W = \sin^{-1}\sqrt{0.225}$, the following results are obtained as theoretical limits,

$$B^{(3)} \underset{g \to 0}{\to} -i \frac{\kappa^2}{B^{(0)}} W_3^{(1)} \times W_3^{(2)},$$

$$B^{(3)} \underset{g' \to 0}{\to} -i \frac{\kappa^2}{B^{(0)}} X^{(1)} \times X^{(2)}.$$
(332)

These limits show that W_3 and X have the units and polarization properties of A, the electromagnetic vector potential. So if A_μ is generalized to $A_\mu^{(1)}$, both W_μ^3 and X_μ must be generalized in the same way, meaning that $W_\mu^{(1)}$ is no longer fixed in isospin axis 3, and that $X_\mu^{(1)}$ is no longer an isospin scalar. These modifications to GWS theory must also be carried out in such a way as to maintain agreement with experimental data. If this is achieved, then $W_\mu^{(1)}$ and $X_\mu^{(1)}$ would become components of $A_\mu^{(1)}$ as follows,

$$A_\mu^{(1)} = x W_{3\mu}^{(1)} + y X_\mu^{(1)},$$
(333)

where x and y are simple scalars, and considerations for $A_\mu^{(1)}$ in isospin space would also be considerations for $W_{3\mu}^{(1)}$ and $X_\mu^{(1)}$.

5.1 SUMMARY OF THE NON-ABELIAN FEATURES OF $W_{3\mu}^{(1)}$ AND $X_\mu^{(1)}$

In this section we summarize the non-Abelian theory of $A_\mu^{(1)}$ in the isospin space (1), (2) and (3). The properties herein summarized for $A^{(1)}$ also apply to $W_{3\mu}^{(1)}$ and to $X_\mu^{(1)}$. The field tensor $G_{\mu\nu}$ is defined in the circular basis from Appendix B by the commutator of covariant derivatives

$$G_{\mu\nu} = \frac{\hbar}{e} [D_\mu, D_\nu],$$
(334)

which reduces to

$$G_{\mu\nu} = [\partial_\mu, A_\nu] + \frac{e}{\hbar} [A_\mu, A_\nu],$$
(335)

(cf. Ryder's [16] Eq. (3.165)). This is a relation between scalar components of the isospin vector $G_{\mu\nu}$, and can be

Summary of Non-Abelian Features of $W_{3\mu}^{(1)}$ and $X_\mu^{(1)}$

expressed as

$$G_{\mu\nu} = \left[\partial_\mu + \frac{e}{\hbar}A_\mu, A_\nu\right]. \tag{336}$$

Therefore the scalar components of $G_{\mu\nu}$ can be built up from the four-curl of A_ν using the O(2) covariant derivative instead of the ordinary derivative ∂_μ, i.e.,

$$\partial_\mu \to \partial_\mu + \frac{e}{\hbar}A_\mu. \tag{337}$$

With the charge quantization condition

$$\partial_\mu = \frac{e}{\hbar}A_\mu, \tag{338}$$

the field tensor $G_{\mu\nu}$ reduces (Appendix D) to the O(2) scalar $F_{\mu\nu}$, the ordinary antisymmetric field tensor of non-Abelian theory. Therefore by replacing ∂_μ of the vacuum O(2) theory of the electromagnetic sector by $\partial_\mu + \frac{e}{\hbar}A_\mu$ we obtain one scalar component of $G_{\mu\nu}$ in O(3) theory.

The transition from O(2) to O(3) theory is completed by adding the isospin indices of O(3), using the circular basis as described in Appendices A and B,

$$G_{\mu\nu}^{(3)*} = \left[\partial_\mu^{(0)}, A_\nu^{(3)*}\right] - i\frac{e}{\hbar}\left[A_\mu^{(1)}, A_\nu^{(2)}\right], \text{ et cyclicum.} \tag{339}$$

Finally, the charge quantization condition (338) is applied to each component,

$$\partial_\mu^{(i)} = \frac{e}{\hbar}A_\mu^{(i)}, \quad (i) = (1), (2), (3), \tag{340}$$

thus identifying $A_\mu^{(i)}$ as momentum operators in the isospin space (1), (2) and (3). The field tensor $G_{\mu\nu}$ is thus identified as the sum of four-curls,

$$G_{\mu\nu}^{(3)*} = \left[\partial_\mu^{(0)}, A_\nu^{(3)*}\right] - i\left[\partial_\mu^{(1)}, A_\nu^{(2)}\right], \tag{341}$$

and becomes similar in structure to the ordinary four-curl definition of $F_{\mu\nu}$ in O(2).

For example, the field $B^{(3)}$, and its dual $-iE^{(3)}/c$ are obtained from XY components and Z4 components of the tensor

Chapter 5. $B^{(3)}$ in Unified Field Theory

$G_{\mu\nu}^{(3)*}$ as follows. Firstly, the magnetic field $B^{(3)}$ is obtained from

$$B_Z^{(3)} = B_Z^{(3)*} = G_{XY}^{(3)*} = \left[\partial_X^{(0)}, A_Y^{(3)*}\right] - i\left[\partial_X^{(1)}, A_Y^{(2)}\right]$$
$$= -i\left(\partial_X^{(1)} A_Y^{(2)} - \partial_Y^{(1)} A_X^{(2)}\right) = -i(\partial^{(1)} \times \mathbf{A}^{(2)})_Z, \quad (342)$$

(because $\mathbf{A}^{(3)*}$ has no Y component by definition of (3) as the Z axis of configuration space). Using Eqs. (340), this definition of $B^{(3)}$ becomes identifiable as a curl of a vector potential, i.e.,

$$B^{(3)*} = \nabla^{(1)} \times \mathbf{A}^{(2)}, \quad \nabla^{(1)} := -i\partial^{(1)}, \quad (343)$$

and the components have been identified as a momentum operator $\nabla^{(1)}$ in the circular state (1), which is also a differential operator of the same circular state. This is a transverse momentum or differential operator. Similarly, reversing the X and Y subscripts reverses the sign of the $B^{(3)}$ component in this definition,

$$B_Z^{(3)} = B_Z^{(3)*} = -G_{YX}^{(3)*} = -i\left(\partial_Y^{(1)} A_X^{(2)} - \partial_X^{(1)} A_Y^{(2)}\right). \quad (344)$$

Using the same procedure for the Z and time component 4 of the four-vectors produces a hypothetical real electric field, defined by its Z component,

$$G_{Z4}^{(3)*} =? -iE_Z^{(3)*} = \left[\partial_Z^{(0)}, A_4^{(3)*}\right] - i\left[\partial_Z^{(1)}, A_4^{(2)}\right]. \quad (345)$$

By definition of $\mathbf{A}^{(1)}$ and $\mathbf{A}^{(2)}$ as transverse plane waves they have no Z or 4 components, however, so the second commutator vanishes, leaving

$$G_{Z4}^{(3)*} = -iE_Z^{(3)*} = \left[\partial_Z^{(0)}, A_4^{(3)*}\right]. \quad (346)$$

However, the (3) component of $A_\mu^{(3)}$ is independent of time and purely irrotational, so the first commutator also vanishes, indicating that there is no real Z axis electric field. This is consistent with the fact that if $B^{(3)}$ is real, then $-iE^{(3)}/c$ is pure imaginary. Other relations of this type are

Summary of Non-Abelian Features of $W_{3\mu}^{(1)}$ and $X_\mu^{(1)}$

summarized in Appendix D.

In general, therefore, the charge quantization condition of Chap. 4 is the scalar part of the vector relations,

$$e\mathbf{A}_\mu^{(1)} = \hbar \partial_\mu^{(1)}, \quad (i) = (1), (2), (3), \qquad (347)$$

which identify photon linear momentum as $e\mathbf{A}_\mu^{(1)}$ in free space. The only non-zero, time-averaged linear momentum component is that in the Z axis, as discussed in Chap. 11 of Vol. 1. These relations give physical meaning to the vector potential in free space as a linear momentum four-vector, which becomes a differential operator using the axiom of quantum mechanics,

$$\mathbf{p}_\mu^{(1)} = \hbar \partial_\mu^{(1)}. \qquad (348)$$

The charge e therefore makes $\mathbf{A}_\mu^{(1)}$ directly proportional to $\mathbf{p}_\mu$ in free space, meaning that electromagnetism in free space is the agent of interaction between two electrons.

Therefore the field tensor of O(3) can be expressed as

$$G_{\mu\nu}^{(3)*} = \left[\partial_\mu^{(0)}, A_\nu^{(3)*}\right] - i\frac{e}{\hbar}\left[A_\mu^{(1)}, A_\nu^{(2)}\right], \qquad (349)$$

where the scalar differential operator is the magnitude of the vector ∂_μ defined in isospin space by

$$\partial_\mu = \partial_\mu^{(1)} + \partial_\mu^{(2)} + \partial_\mu^{(3)}. \qquad (350)$$

Therefore there also exist cyclic relations such as

$$\partial_\mu^{(1)} \times \partial_\nu^{(2)} = i\partial_\mu^{(0)} \partial_\nu^{(3)*}, \qquad (351)$$

between the various differential, or momentum operators.

The essential outcome of these considerations therefore is that $\mathbf{A}_\mu^{(1)}$ couples to the field momentum $\mathbf{p}_\mu^{(1)}$ through the charge e. Therefore $\mathbf{A}_\mu^{(1)}$ becomes physically meaningful as a field momentum four-vector, which in the quantum theory is a photon momentum. Therefore, in GWS theory, $W_{3\mu}^{(1)}$ and $X_\mu^{(1)}$ become components of this photon momentum. These considerations flow naturally from the fact that if $\mathbf{A}^{(1)}$ and $\mathbf{A}^{(2)}$ are defined as O(2) plane waves, the vector cross product

Chapter 5. $B^{(3)}$ in Unified Field Theory

$\mathbf{A}^{(1)} \times \mathbf{A}^{(2)}$ is non-zero, and proportional to $\mathbf{B}^{(3)}$, a physical magnetic field *orthogonal* to the plane of definition in O(2) theory. Taking isospin components as (1), (2) and (3) of the circular basis of configuration space, we define the covariant derivative D_μ in terms of $\mathbf{A}^{(1)} \times \mathbf{A}^{(2)}$ and also the field tensor $G_{\mu\nu}$ in terms of

$$G_{\mu\nu} = \left[\partial_\mu + \frac{e}{\hbar}A_\mu, A_\nu\right] = \frac{\hbar}{e}[D_\mu, D_\nu]. \tag{352}$$

These procedures define the field tensor $G_{\mu\nu}$ as the four-curl from the covariant derivative of A_ν in O(2), or as the commutator $[D_\mu, D_\nu]$ of O(3). Since $A_\mu^{(1)} = xW_{3\mu}^{(1)} + yX_\mu^{(1)}$ it is clear that $W_{3\mu}^{(1)}$ and $X_\mu^{(1)}$ must follow the analysis for $A_\mu^{(1)}$, the proportions x and y being fixed by experiment. From the algebraic relations

$$\mathbf{A}^{(1)} = \frac{\mathbf{B}^{(1)}}{\kappa} = c\frac{\mathbf{B}^{(1)}}{\omega}, \qquad \mathbf{A}^{(2)} = \frac{\mathbf{B}^{(2)}}{\kappa} = c\frac{\mathbf{B}^{(2)}}{\omega}, \tag{353}$$

it becomes clear that $W_{3\mu}^{(1)}$ and $X_\mu^{(1)}$ are magnetic fields which are also subjected to the link between O(2) and O(3) theory. This link is forged by replacing ∂_μ of O(2) theory by the O(2) covariant derivative $\partial_\mu + (e/\hbar)A_\mu$, which acts in O(3) through Eq. (B12) of Appendix B. These procedures take place in free space electromagnetism. Finally in this section, we can write

$$G_{\mu\nu}^{(3)*} = \partial_\mu^{(0)} \times \mathbf{A}_\nu^{(3)*} - i\partial_\mu^{(1)} \times \mathbf{A}_\nu^{(2)}, \tag{354}$$

and for space indices, $\partial_\mu^{(0)} \times \mathbf{A}_\nu^{(3)*}$ becomes identifiable as the ordinary curl $\nabla \times \mathbf{A}^{(3)*}$ of O(2). Thus $\partial_\mu^{(1)} \times \mathbf{A}_\nu^{(2)}$ in this context is also a type of curl operation in O(3). It is not considered, of course, in O(2), because $\mathbf{A}^{(1)} \times \mathbf{A}^{(2)}$ is not considered conventionally. However, it is non-zero in O(2) as well as in O(3), and as we have seen, self-indicates the existence of $\mathbf{B}^{(3)}$. The latter mediates physical effects of magnetization by light, and so we conclude that O(2) theory self-indicates the need for O(3) theory.

Therefore, the SU(2) ⊗ U(1) product group symmetry of GWS must also be expanded to take account of the expansion of its electromagnetic sector from U(1) to O(3).

5.2 SPECIFIC EFFECTS OF $B^{(3)}$ IN GWS THEORY

In this section it is shown that the existence of $B^{(3)}$ in vacuum electromagnetism provides the additional inference that there exist three circular polarization states ((1), (2) and (3)) for the vector bosons W_3 and X. The major features of GWS theory are maintained intact in the presence of $B^{(3)}$, and the latter does not affect the ability of GWS to produce the experimentally observed boson masses and characteristic $\hat{P}$ violating effects which have been detected [60]. It is straightforward to work $B^{(3)}$ into GWS theory provided that a careful distinction be maintained between the abstract isospin space (1,2,3) and the physical frame ((1), (2), (3)) used in Chaps. 3 and 4 for the electromagnetic potential $A_\mu^{(i)}$ in O(3) theory. These two frames and spaces are not the same. This is readily inferred from the fact that $W_{3\mu}$ is a four-vector and proportional (Sec. 5.1) to A_μ of the U(1) sector. Thus, in an O(3) theory of electromagnetism, we obtain components such as $W_3^{(1)}$, $W_3^{(2)}$, and $W_3^{(3)}$, i.e., the isospin index 3 becomes associated with three space indices (1), (2) and (3). In this sense, the U(1) sector of GWS is enlarged to O(3). Since $W_{3\mu}$ and X_μ are parts of A_μ in GWS, they are plane waves if A_μ contains circular components (1) and (2) which are also plane waves,

$$W_3^{(1)} = \frac{W_3^{(0)}}{\sqrt{2}}(i\mathbf{i} + \mathbf{j})e^{i\phi}, \qquad W_3^{(2)} = \frac{W_3^{(0)}}{\sqrt{2}}(-i\mathbf{i} + \mathbf{j})e^{-i\phi}, \qquad (355)$$

and

$$W_3^{(1)} \times W_3^{(2)} = iW_3^{(0)} W_3^{(3)*} = -W_3^{(0)}\left(iW_3^{(3)}\right)^*, \qquad (356)$$

with cyclic permutations. Equation (356) is directly analogous with

$$\mathbf{A}^{(1)} \times \mathbf{A}^{(2)} = -A^{(0)}(i\mathbf{A}^{(3)})^*, \qquad (357)$$

with cyclic permutations, so that if $W_3^{(1)}$ and $W_3^{(2)}$ are polar vectors, $iW_3^{(3)}$ is an axial vector. Similarly,

$$X^{(1)} = \frac{X^{(0)}}{\sqrt{2}}(i\mathbf{i} + \mathbf{j})e^{i\phi}, \qquad X^{(2)} = \frac{X^{(0)}}{\sqrt{2}}(-i\mathbf{i} + \mathbf{j})e^{-i\phi}, \qquad (358)$$

and

$$\boldsymbol{X}^{(1)} \times \boldsymbol{X}^{(2)} = -X^{(0)}(i\boldsymbol{X}^{(3)})^*, \qquad (359)$$

with cyclic permutations. Thus, both $W_3^{(1)}$ and $\boldsymbol{X}^{(1)}$ are described by O(3) gauge geometry, as for $\boldsymbol{A}^{(1)}$. As in Chaps. 3 and 4, the conjugate product $\boldsymbol{A}^{(1)} \times \boldsymbol{A}^{(2)}$ is an intrinsic part of O(3) gauge geometry in an isospin space which is also the configuration space in the circular basis (1), (2) and (3). Before proceeding to a more detailed discussion of the effect on GWS of the novel vector boson components $W_3^{(3)}$ and $\boldsymbol{X}^{(3)}$, a brief review is given of the key aspects of unified field theory [16,47,60]. In GWS, weak isospin (I_W) is locally conserved through a scalar interaction between the isospinor L, and a boson W in weak isospin space, a scalar interaction of the type [47]

$$gL^+\boldsymbol{\sigma}L \cdot \boldsymbol{W} = \frac{g}{2}[\overline{\nu}_e\ \overline{e}_L]\begin{bmatrix} W_3 & W_1 - iW_2 \\ W_1 + iW_2 & -W_3 \end{bmatrix} = \frac{g}{2}\overline{\nu}_e W_3 \nu_e$$
$$-\frac{g}{2}\overline{e}_L W_3 e_L + \frac{g}{\sqrt{2}}\overline{\nu}_e W_+ e_L + \frac{g}{\sqrt{2}}\overline{e}_L W_- \nu_e, \qquad (360)$$

where

$$W_\pm := \frac{1}{\sqrt{2}}(W_1 \mp iW_2). \qquad (361)$$

Equation (360) shows that in this view, the interaction between $\overline{\nu}_e$ and ν_e (i.e., the neutrino interaction) is mediated by the same boson W_3 as that between the electrons. Therefore, in GWS, $W_{3\mu}$ cannot be identified directly with A_μ, because the neutrino does not interact with the electromagnetic field. The neutral weak boson is introduced therefore to ensure that the interaction of the electron neutrino (ν_e) with the unified (electroweak) field is different from that of the electron. In order for this to be so, X_μ must be a scalar in the isospin space (1, 2, 3) and must interact with an isoscalar L^+L, where $L^+ := [\overline{\nu}_e,\ \overline{e}_L]$ is the hermitian conjugate of L defined in Eq. (321). The structure of GWS depends specifically on the fact that W_μ is a vector and X_μ is a scalar in the isospin space (1, 2, 3). The introduction of $B^{(3)}$ into GWS must be done in such a way as to conserve this key feature. If W_μ were made an isovector, for example,

Specific Effects of $B^{(3)}$ in GWS Theory

we would again obtain the result that the neutrino and electron would interact with the unified electroweak field, this time through the same combination of bosons (components of W_μ and the putative X_μ) and this is not physically meaningful. The isoscalar nature of X_μ cannot therefore be changed to that of a vector in the same abstract isospin space (1, 2, 3).

Since L^+L is an isoscalar, GWS has the additional field particle interaction [47],

$$\frac{g'}{2}L^+LX_\mu = \frac{g'}{2}[\overline{v}_e \ \overline{e}_L]\begin{bmatrix}v_e\\e_L\end{bmatrix} = \frac{g'}{2}(\overline{v}X_\mu v + \overline{e}X_\mu e). \qquad (362)$$

The combined interaction in GWS is therefore, in simplified terms [47],

$$\frac{g'}{2}L^+LX_\mu - \frac{g}{2}L^+\sigma L \cdot W_\mu = \frac{1}{2}\overline{v}_e(g'X_\mu - gW_{3\mu})v_e$$
$$+ \frac{1}{2}\overline{e}_L(g'X_\mu + gW_{3\mu})e_L - \frac{g}{\sqrt{2}}(\overline{v}_e W_{+\mu}e_L + \overline{e}_L W_{-\mu}v_e). \qquad (363)$$

A more detailed description of the theory is given by Ryder [16], but for our present purposes we note from Eq. (363) that electromagnetism can be identified in this simplified description [47] as

$$A_\mu = g'X_\mu + gW_{3\mu}, \qquad (364)$$

and more properly [16] as the normalized Eq. (327), using the Weinberg angle θ_W. The conserved quantity associated with W_μ is weak hypercharge, Y^W, to which the gauge field W_μ is coupled. Similarly, electric charge, Q, couples to the electromagnetic gauge A_μ, and weak isospin, I^W, to the vector boson $W_\mu = (W_{+\mu}, W_{3\mu}, W_{-\mu})$. The conserved quantities are related by

$$Q = I_3^W + \frac{Y^W}{2}, \qquad (365)$$

in analogy with the Gell-Mann-Nishijima relation of strong force theory [47], where Y_S^W, the analogue of Y^W, is given for the strong force by B + S, where B is baryon number and S is strangeness. The physical bosons in GWS are therefore

A_μ and Z_μ, given by

$$Z_\mu = W_{3\mu} \cos\theta_W - X_\mu \sin\theta_W, \quad (366a)$$

a four-vector orthogonal to A_μ. The physical bosons A_μ and Z_μ are both weighted mixtures of $W_{3\mu}$ and X_μ.

The GWS model is completed [16,47,60] by a model mechanism of non-Abelian SSB, whereby the boson Z_μ acquires mass, but A_μ is left massless. The neutrino couples to the Z_μ field only, and not to the electromagnetic field, but the left handed part of the electron couples to both A_μ and the weak Z_μ field, introducing parity violating effects into atomic and molecular spectroscopy [60]. The hypothesis of I^W and Y^W conservation therefore necessitates the introduction of four gauge fields, A_μ, Z_μ, $W_{+\mu}$ and $W_{-\mu}$; the field Z_μ gives rise to neutral current processes which have been observed experimentally. The W_μ and Z_μ bosons have also been observed at the predicted masses, provided that Higgs SSB is incorporated in a well-defined but delicately modelled way.

With the advent of $B^{(3)}$ in the electromagnetic field, we have seen that the $W_{3\mu}$ and Z_μ bosons acquire three states of circular polarization, (1), (2) and (3). The mass of Z_μ is determined in GWS by the premultiplier in the relevant Lagrangian of the term $(\eta^2/4)(gW_{3\mu} - g'X_\mu)^2$ [16]. Expanding the $W_{3\mu}W_{3\mu}$ term, for example,

$$W_{3\mu}W_{3\mu} = \mathbf{W}_3^{(1)} \cdot \mathbf{W}_3^{(1)*} + \mathbf{W}_3^{(2)} \cdot \mathbf{W}_3^{(2)*} + \mathbf{W}_3^{(3)} \cdot \mathbf{W}_3^{(3)*} - W_3^{(0)2}, \quad (366b)$$

which contains the additional $\mathbf{W}_3^{(3)} \cdot \mathbf{W}_3^{(3)*} - W_3^{(0)2}$, analogous with the additional $\mathbf{A}^{(3)} \cdot \mathbf{A}^{(3)*} - A^{(0)2}$ in electromagnetism. From Appendix D, however, it is clear that

$$\mathbf{A}^{(3)} \cdot \mathbf{A}^{(3)*} - A^{(0)2} = 0, \quad (367)$$

because $|\mathbf{A}^{(3)}| = A^{(0)}$. Therefore $B^{(3)}$ makes no difference to $W_{3\mu}W_{3\mu}$ or to $X_\mu X_\mu$, and from Eqs., (327) and (367), it makes no difference to the cross term $W_{3\mu}X_\mu$. It would therefore appear that $B^{(3)}$ makes no difference to the observed masses of Z_μ and $W_{\pm\mu}$, unless the premultiplier of transverse terms such as $\mathbf{W}_3^{(1)} \cdot \mathbf{W}_3^{(1)*} + \mathbf{W}_3^{(2)} \cdot \mathbf{W}_3^{(2)*}$ were for some reason different from that of the longitudinal terms such as $\mathbf{W}_3^{(3)} \cdot \mathbf{W}_3^{(3)*} - W_3^{(0)2}$. This does not appear, however, to be very likely, because mass is a

Specific Effects of $B^{(3)}$ in GWS Theory

scalar Lorentz invariant. Four-vector properties such as $A_\mu^{(i)} A_\mu^{(i)}$, $W_{3\mu}^{(i)} W_{3\mu}^{(i)}$ and so on are also Lorentz invariants, and can appear as terms in a Lagrangian provided they are premultiplied by a scalar, a term that indicates [16] the presence of mass. Thus, the premultiplier of $Z_\mu Z_\mu$ in the GWS Lagrangian indicates that Z_μ has mass, a mass that is relativistically invariant and thus cannot have different transverse or longitudinal components.

It is clear in summary that $B^{(3)}$ does not affect the ability of GWS to predict the correct masses of Z_μ and $W_{+\mu}$, but leads to the novel polarization state (3) for massive bosons, $Z_\mu^{(3)}$, $W_{3\mu}^{(3)}$, $X_\mu^{(3)}$ and $X_{+\mu}^{(3)}$. The states $\boldsymbol{X}^{(3)}$ and $\boldsymbol{W}_3^{(3)}$ contribute to $\boldsymbol{B}^{(3)}$, and are therefore observed in phenomena of magnetization by electromagnetic radiation, for example at microwave frequencies.

5.3 SSB AND PHOTON MASS IN GWS

In Vol. 1, and in previous chapters of this volume, we have argued that longitudinal photon polarization, indicated by the existence of the physical $\boldsymbol{B}^{(3)}$ field of light, means that the photon cannot be massless, because a massless boson has two degrees of polarization in the vacuum. This result is derived from special relativity [22], as first demonstrated by Wigner. In a boson, these states correspond to angular momentum eigenvalues $\pm \hbar$, i.e., to states with quantum number $J = \pm 1$. The state 0 is disallowed because the particle travels at the speed of light, and this leads to the result that the little group [16,22] is E(2), the group of rotations and translations in a plane. The E(2) group is, however, *unphysical*, and this means that the existence of a massless particle travelling always at the speed of light in all frames of reference *is also unphysical*. As soon as the particle (our boson) acquires mass, however tiny, then $J = 0, \pm 1$, and there are three states of physical polarization in the vacuum. The anomaly of the E(2) group is removed, and the particle becomes physical and relativistic, the range of electromagnetic radiation becomes finite, meaning that its intensity diminishes with distance. Therefore a photon with mass has three states of helicity, +1, 0, -1. In the Higgs mechanism discussed in Chap. 2, the extra helicity state 0 is obtained from a theoretically massless scalar particle, i.e., from the Higgs field, and in so doing, the photon acquires mass simultaneously [16].

In conventional gauge theory, however, *bare* photon mass

Chapter 5. $B^{(3)}$ in Unified Field Theory

is disallowed in the absence of SSB (i.e., in the absence of a degenerate vacuum) because a mass term $m_0 A_\mu A_\mu$ is not gauge invariant conventionally. In Vol. 1 and previous chapters of this volume we have introduced the condition

$$A_\mu A_\mu = 0, \quad m_0 \neq 0, \tag{368a}$$

which is equivalent to the conventional

$$A_\mu A_\mu \neq 0, \quad m_0 = 0. \tag{368b}$$

Condition (368a) means that [8] A_μ becomes a light-like four-vector, and $m_0 \neq 0$ becomes compatible with gauge invariance. This, in turn, unifies two lines of thought in contemporary field theory: 1) gauge invariance; 2) electromagnetic theory with non-zero photon mass. Equation (368a), however, is not compatible with a transverse, or Coulomb, gauge, in which the vector potential is

$$\mathbf{A}^{(1)} = \mathbf{A}^{(2)*} = \frac{A^{(0)}}{\sqrt{2}}(i\mathbf{i} + \mathbf{j})e^{-i\phi}, \tag{369}$$

and in which (Appendices D and E)

$$A_\mu^{(1)} = A_\mu^{(2)*} = (\mathbf{A}^{(1)}, 0), \quad A_\mu^{(3)} = i(\mathbf{A}^{(3)}, iA^{(0)}). \tag{370}$$

From Eq. (370), it is found that

$$A_\mu A_\mu = A_\mu^{(1)} A_\mu^{(1)*} + A_\mu^{(2)} A_\mu^{(2)*} + A_\mu^{(3)} A_\mu^{(3)*} \neq 0, \tag{371}$$

but that

$$A_\mu^{(3)} A_\mu^{(3)*} = 0. \tag{372}$$

Equation (372), as shown in Appendices D and E, is a result of O(3) gauge geometry applied to free space electromagnetism, *provided that* Eq. (369) is accepted for $A_\mu^{(1)}$ and $A_\mu^{(2)*}$, and therefore for $\mathbf{B}^{(1)}$, $\mathbf{B}^{(2)}$, $\mathbf{E}^{(1)}$, and $\mathbf{E}^{(2)}$. These are, of course, the conventional plane waves of vacuum electrodynamics with O(2) gauge geometry.

As we have seen, these transverse, O(2) fields indicate

SSB and Photon Mass in GWS

the existence of $B^{(3)}$ in the vacuum and therefore the need to extend O(2) to O(3) as in the two previous chapters. The existence of $B^{(3)}$ in turn leads to the inference of non-zero photon mass, because the photon develops a third degree of polarization represented by the concomitant physical field $B^{(3)}$. This (3) state does not exist self-consistently in flat, O(2), gauge geometry. Finally, in order to make photon mass consistent with gauge invariance, we must either: a) develop an O(3) potential model fully consistent with Eq. (368a); or b) provide the photon with mass through a Higgs mechanism as discussed in Chap. 2. Choice (a) leads to the abandonment of the Coulomb gauge (369), and step (b) means the replacement of the d'Alembert equation with the Proca equation as discussed in Chap. 2.

Since photon mass is experimentally $\leq (10^{-45}\text{-}10^{-65})$ kgm, the Proca equation gives a $B^{(3)}$ (Chap. 2) which is practically indistinguishable in laboratory experiments from that from the d'Alembert equation, so that it is plausible to proceed through a *perturbation* of the d'Alembert equation with a Higgs mechanism that leads as in Chap. 2 to a Proca equation and slowly decaying fields $B^{(1)}$, $B^{(2)}$ and $B^{(3)}$ with finite range. This means effectively a perturbation of the useful Coulomb gauge instead of its abandonment. The task in GWS then becomes one of incorporating the Higgs field into the model in such a way as to give the experimental photon mass self consistently with those of the bosons Z_μ and $W_{\pm\mu}$. A substantial amount of work is available on this problem and the reader is referred to papers by Huang [35] and references therein. For our purposes, suffice it to mention that GWS is based on delicate modelling [16], because if Z_μ, $W_{\pm\mu}$, $W_{3\mu}$, and X_μ are to be *gauge* fields, they must also be gauge invariant in the presence of their own mass.

5.3.1 SSB AS THE SOURCE OF PHOTON MASS IN ABELIAN THEORY

By considering SSB of the Abelian Lagrangian (197) of Chap. 2, the result is obtained [16] that the photon becomes massive, and the Lagrangian acquires a term proportional to $A_\mu A_\mu$. This is the Higgs phenomenon and is the result of a particular model of the vacuum itself as discussed in Chap. 2. The originally two dimensional, massless, photon becomes a three dimensional massive boson by picking up a third degree of freedom from a Higgs field. Such a result is obtained from a Lagrangian, Eq. (197), which is compatible with gauge invariance, and so the spontaneous symmetry breaking of Abelian gauge geometry is applied to obtain a non-zero photon mass in a manner that is compatible with gauge invariance of the second kind. The Higgs mechanism is therefore capable of removing the problem discussed in Sec. 5.3, i.e., if the vacuum symmetry is broken, as discussed in Chap. 2, a term $m_0 A_\mu A_\mu$ in the Lagrangian becomes compatible with gauge invariance requirements. Such a term leads to the replacement of the d'Alembert with a Proca equation as discussed in Chap. 2. As mentioned by Ryder [16] on his page 301 "...the photon has eaten a scalar field and has acquired a mass." *The existence of three degrees of polarization for the photon is precisely what is indicated by $B^{(3)}$, which is therefore an indicator of photon mass.* The Higgs mechanism allows this development to be compatible with $A_\mu A_\mu \neq 0$ as indicated by Appendix D, while also retaining compatibility with gauge invariance. Therefore it is no longer consistent to assert that the photon mass must be zero, and in view of the unphysical nature of E(2), never has been. It is important to note that $B^{(3)}$ is non-zero for identically zero photon mass, but its very existence means that the photon has three degrees of polarization, thereby indicating that its mass, for self-consistency, must be physically non-zero. The fundamental reason for this is in the Wigner paper of 1939 [22] that first indicated the fact that the little group is unphysical for any particle that has no mass. A flat particle (with two degrees of polarization) is not a physically meaningful entity. As soon as it acquires mass, however, it simultaneously acquires the necessary third axis, and, if it is a boson, three helicities as argued already. The extra one, 0, is obtained in the Higgs mechanism by breaking spontaneously the vacuum symmetry. Clearly therefore, $B^{(3)}$ in the vacuum means a finite photon mass in the vacuum. In this respect, we differ from Ryder [16,47], who insists on a rigorously zero photon mass. As Ryder himself shows, however, SSB of an initially gauge invariant theory leads to the inference that photon mass in the vacuum with broken symmetry is *non*-zero. Such a result is obtained from a theory that is originally compatible with gauge invariance of the second kind. We are driven to conclude that either the Higgs mechanism is itself incompatible with gauge invariance, or that the photon mass in the vacuum is non-zero. We choose the second option because the Higgs mechanism is a key ingredient of GWS, and is the basis of GWS to predict correctly the observed masses of Z_μ and $W_{\pm\mu}$. If the Higgs mechanism is thus compatible with experimental data, the argument cannot then be sustained that it is incompatible with a fundamental theoretical principle such as gauge

SSB and Photon Mass in GWS

invariance. Indeed, Z_μ and $W_{\pm\mu}$ are themselves gauge fields, and are at the same time renormalizable massive fields. Proceeding in this way, the theory of finite photon mass has been incorporated [35] in GWS and SU(5).

5.3.2 SSB AS THE SOURCE OF PHOTON MASS IN NON-ABELIAN THEORY

SSB of O(3) gauge geometry was originally discussed by Kibble [61] and shows that two out of the three components of the vector $\mathbf{A}_\mu^{(1)}$ become massive. This result is firmly rooted in fundamental group theory. If, in the abstract isospin space, 1, 2 and 3 are the isospin indices then components 1 and 2 become massive while 3 remains massless. This has interesting analogies with the mechanism outlined in Sec. 5.3. In this model one massless field remains because the subgroup, U(1), of O(3) under which the vacuum remains invariant has only one generator. In GWS, the SSB of a non-Abelian gauge geometry leads to two massive bosons and one massless boson. The isospin indices in this case are clearly those of the abstract space, and not the indices (1), (2) and (3) of the two previous chapters.

Isospin, or isotopic spin, invariance is described by the same group of rotations in three dimensions (SU(2) or O(3)), but isospin space (1, 2, 3) is a purely abstract concept, based originally [60] on the assertion that the proton and neutron are two states of a single particle, the nucleon, N, with spin $I = 1/2$. The two component states with I_3, along axis 3, are given by $I = \pm 1/2$ and are the proton and neutron respectively. Electric charge is related to I_3 through

$$Q = I_3 + \frac{1}{2}. \tag{373}$$

The space (1), (2), (3) on the other hand is the circular representation of 3-D configuration space, the physical space of rotations under which angular momentum is invariant. The isospin space 1, 2, 3 is a purely abstract space, whose third axis is related to charge, Q, through Eq. (373). The two spaces are, however, governed by the same rotation group, O(3), and the same gauge geometry. The massive bosons $W_{\pm\mu}$ and Z_μ are physically different bosons, with different masses, but (1), (2) and (3) must be components in the circular basis of the same boson, e.g. a photon. Thus, components (1) and (2) cannot have a different mass from component (3); whereas SSB of a non-Abelian gauge geometry in space (1,2,3) produces two massive bosons and one massless. Therefore GWS is based on a modelling procedure which results in zero photon mass identically. On the other hand, SSB of Abelian gauge geometry produces a single massive boson as in Sec. 5.3.1. With the advent of $B^{(3)}$, the assertion of identically zero photon mass becomes untenable, because $B^{(3)}$ signals the presence of a physical third axis. The well known Wigner theory [22] asserts on the other hand that there can be no physical third axis in a massless particle. It is clear that an additional mechanism of SSB must be incorporated in GWS to produce a non-zero photon mass, since it is illogical to assert that $B^{(3)}$ is zero or otherwise unphysical. This mechanism can be applied to A_μ from the final result of GWS (which agrees with experimental data) to provide it with the appropriate photon mass. Alternatively, a mechanism such as that developed by Huang [35] can be used to attempt to produce the photon mass self-consistently with those of $W_{\pm\mu}$ and Z_μ without affecting the ability of GWS to produce the experimentally observed masses for the latter.

Chapter 6. $B^{(3)}$ in Quantum Electrodynamics

The emergence of $B^{(3)}$ in classical electrodynamics means that it has its counterpart in quantum field theory, referred to in Vol. 1 as the photomagneton, $\hat{B}^{(3)}$, a field operator [1-10]. This concept was developed in Chap. 3 of that volume in terms of simple Schrödinger equations, and it was demonstrated that $B^{(3)}$ is a well defined expectation value in quantum mechanics. The development of $\hat{B}^{(3)}$ in quantum electrodynamics (QED) is an interesting procedure because it is necessary to demonstrate that it does not affect the ability of QED to produce results such as the anomalous magnetic moment of the electron to several decimal places. More fundamentally, $B^{(3)}$ in classical, Abelian and non-Abelian electrodynamics must be shown to be compatible with renormalization in QED. In this chapter, it is shown that the classical $B^{(3)}$ leaves the structure of QED unaffected, a result which is of course consistent with Chap. 1 of this volume, where it was demonstrated that $B^{(3)}$ is a direct result of the Dirac equation describing intrinsic electron spin in a classical electromagnetic field. This "semi-classical" result (quantized spin, classical field) is consistent with the classical result of Chap. 12 of Vol. 1, where $B^{(3)}$ emerged from the relativistic Hamilton-Jacobi equation of an electron as classical charged particle in the classical field. It is therefore expected that QED produce $\hat{B}^{(3)}$ from a consideration of the quantized electron in the quantized field.

6.1 CANONICAL QUANTIZATION AND $B^{(3)}$

As discussed in Chap. 10 of Vol. 1, canonical quantization of the massless electromagnetic field is beset with difficulty, and relies on the usual assumption that there are only two (transverse) degrees of physical polarization. The gauge geometry in this view is the flat O(2), in which the conjugate product $A^{(1)} \times A^{(2)}$ is asserted to be zero. This assumption is contradicted experimentally as discussed throughout these volumes, and is untenable. The cross product $\mathbf{A}^{(1)} \times \mathbf{A}^{(2)}$ produces a physical field $\mathbf{B}^{(3)}$ in the axis orthogonal to the plane of definition of O(2) symmetry, and this means inevitably that the field after canonical quantization must produce a particle, the photon, with three helicities, 1, 0, and -1. From fundamental special relativity, this conclusion implies in turn that such a three dimensional particle must have mass, which is picked up from a Higgs particle as discussed in the previous chapter. The d'Alembert equation of classical electrodynamics is changed to a Proca equation, which as discussed in Chap. 2 of that volume is the result of spontaneous symmetry breaking of the vacuum in Abelian field theory which is originally compatible with gauge invariance. In other words, as discussed in the previous chapter, the Higgs mechanism produces a massive photon within the framework of an originally O(2) theory, but imbues it simultaneously with an additional degree of polarization.

With these considerations, canonical quantization should be based on the Euler-Lagrange equation of the massive electromagnetic field, with Lagrangian

$$\mathcal{L} = -\frac{1}{4} F_{\mu\nu} F^*_{\mu\nu} - \frac{1}{2} \xi^2 A_\mu A_\mu, \tag{374}$$

where $\xi = m_0 c/\hbar$, m_0 being the photon mass. The canonical momentum, π_μ, (Vol. 1), is well defined in this view because there are three, well-defined, axes of space polarization, and is given by [16]

$$\pi_\mu = \frac{\partial \mathcal{L}}{\partial \dot{A}_\mu} := \frac{\partial \mathcal{L}}{\partial \left(\frac{\partial A_\mu}{\partial x_0} \right)} := \frac{\partial \mathcal{L}}{\partial (\partial_0 A_\mu)}. \tag{375}$$

Using

$$F_{\mu\nu} = \partial_\mu A_\nu - \partial_\nu A_\mu, \tag{376}$$

it is found that

$$\pi_\mu = \partial_\mu A_0 - \partial_0 A_\mu = \partial_\mu A_0 - \dot{A}_\mu, \tag{377a}$$

so that

Canonical Quantization and $B^{(3)}$

$$\pi_0 = 0, \qquad \pi_i = \partial_i A_0 - \dot{A}_i \tag{377b}$$

where i denotes space axes. Therefore the time-like component of π_μ vanishes. From Appendix D, we have the results

$$A_\mu^{(1)} = (\mathbf{A}^{(1)}, 0) = A_\mu^{(2)*}, \qquad A_\mu^{(3)} = i(\mathbf{A}^{(3)}, iA^{(0)}), \tag{378}$$

from a theory which considers three space polarizations ((1), (2) and (3)), so the term $\partial_i A_0$ in Eq. (377) vanishes, giving the result

$$\pi_i = -\dot{A}_i, \tag{379}$$

or, in vector notation, the familiar result of classical electrodynamics

$$\boldsymbol{\pi} = -\frac{\partial \mathbf{A}}{\partial t} = \mathbf{E}. \tag{380}$$

The momentum classically conjugate to $\mathbf{A}$ is therefore the electric field $\mathbf{E}$. Since $i\mathbf{A}^{(3)}$ in Eq. (378) is rigorously imaginary, divergentless and irrotational,(Appendix D), its real conjugate momentum is zero. This is consistent with the fact that $\mathbf{A}^{(1)} \times \mathbf{A}^{(2)}$ produces a magnetic field $\mathbf{B}^{(3)}$, an axial vector, whereas the real (i.e., physical) electric field is a polar vector. The field $-i\mathbf{E}^{(3)}/c$ is formally dual to the real and physical $\mathbf{B}^{(3)}$, as argued throughout these volumes, but has no real part, and no physical effect at first order.

Canonical quantization proceeds in this view through the usual Heisenberg commutators of the field, as ably described by Ryder [16]. Our purpose here is to introduce the subject of $\mathbf{B}^{(3)}$ in QED by illustrating the fact that the massive classical field, not surprisingly, produces a well-defined massive photon with three degrees of polarization. The existence of these three polarizations is best illustrated through the existence of the by now familiar cyclic relations

$$\mathbf{B}^{(1)} \times \mathbf{B}^{(2)} = iB^{(0)}\mathbf{B}^{(3)*}, \quad \text{et cyclicum}, \tag{381}$$

between three physical, magnetic fields in the vacuum, fields which are mutually orthogonal in the circular basis. The covariant Heisenberg commutator [16] is

$$[\dot{A}_i(\mathbf{x}, t), A_j(\mathbf{x'}, t)] = ig_{ij}\delta^3(\mathbf{x} - \mathbf{x'}), \qquad (382)$$

and vanishes if A_i is time independent. Thus, the longitudinal $i\mathbf{A}^{(3)}$ is unphysical at first order, and plays no part in canonical quantization of the type (382). This does *not* mean that $B^{(3)}$ cannot be quantized, its quantum counterpart is well defined [1-10] as being proportional directly to the longitudinal angular momentum of the photon, i.e., its angular momentum about the Z axis of beam propagation. The expectation value of this angular momentum is the Dirac constant $\hbar$, and the photomagneton operator is

$$\hat{B}^{(3)} = B^{(0)} \frac{\hat{J}}{\hbar}. \qquad (383)$$

The Hamiltonian obtained on canonical quantization of the massive electromagnetic field is proportional to a quadratic product of annihilation and creation operators in all three polarization states, i.e., to $\sum_{\lambda=1}^{3} a^{(\lambda)+} a^{(\lambda)}$. Here, the longitudinal $i\mathbf{A}^{(3)}$ acts at second order, i.e., the product $i\mathbf{A}^{(3)} \cdot (i\mathbf{A}^{(3)})^*$ is real and positive, and therefore physical. The third axis therefore contributes to the Hamiltonian, but does not contribute to the commutator (382). Note carefully that these inferences are based on a particular model, discussed in Appendix D, and summarized in Eqs. (378). This model was developed from a classical, non-Abelian, description of the field, in the massless limit $m_0 \to 0$; and used as an illustration of canonical quantization features when there are three degrees of polarization taken into consideration. More consistently, the Proca equation should be solved as discussed in Chap. 2, and canonical quantization developed of the exponentially decaying classical fields of Eqs. (187) and (188), leading to their description in terms of creation and annihilation operators.

6.2 THE EFFECT OF $B^{(3)}$ ON RENORMALIZABILITY IN QED

In this and in following sections we indicate without unnecessary detail that $B^{(3)}$ does not affect some powerful results of QED, such as its ability to describe very accurately the anomalous magnetic moment of the electron. It is first necessary to prove that the classical $B^{(3)}$ does not destroy renormalizability in QED. This is straightforward,

The Effect of $B^{(3)}$ on Renormalizability in QED

because, following Ryder [16], the general formula for the degree of divergence of a Feynman graph is unaffected by $B^{(3)}$. This formula is, in Abelian QED [16],

$$D = dL - 2P_i - E_i, \qquad (384)$$

where d is the dimension of space-time, L is the number of loops, P_i is the number of internal photon lines, and E_i the number of internal electron lines. If n is the number of vertices, P_e the number of external photon lines, and E_e the number of external electron lines, it can be shown [16] that

$$D = d + n\left(\frac{d}{2} - 2\right) - \left(\frac{d-1}{2}\right)E_e - \left(\frac{d-2}{2}\right)P_e. \qquad (385)$$

When the dimension of space-time, d, is four, the dependence of D on n, the number of vertices, disappears. Clearly, the theory of $B^{(3)}$ in classical electrodynamics is worked out in four-dimensional space-time, as for the usual transverse fields, and maintains the renormalizability of Abelian QED. In particular, the photon self-energy diagram, which has no classical counterpart, is unaffected by the belated recognition of the vacuum $B^{(3)}$ in the classical theory, provided that the overall gauge geometry is maintained at O(2). A more self-consistent analysis requires, as we have argued in Chaps. 3 and 4, that the gauge geometry be extended to O(3), which theory is again renormalizable to all orders in QED [16]. This was discussed briefly in Chap. 4.

In Abelian QED [16], the calculation of the three primitive divergences is carried out using dimensional regularization, which has the effect of multiplying e in the photon/electron Lagrangian by the factor $\mu^{2-d/2}$, where μ is an arbitrary mass and d is the mass dimension [16] of the Lagrangian. This extension to d dimensions in QED is made only for internal loops, and leads via the two parameter Feynman formula to a convergent term denoted $\Lambda_\mu^{(2)}$ in explicit expressions for the three primitively divergent Feynman diagrams. It is this convergent term that gives the anomalous value of the magnetic moment of the electron to several decimal places. The field $B^{(3)}$ has no specific influence on the calculation of $\Lambda_\mu^{(2)}$ in QED, and therefore has no influence on the precisely measured value of the magnetic moment. These points are developed in the following section.

6.3 $B^{(3)}$ AND THE ELECTRON'S MAGNETIC MOMENT

The anomalous magnetic moment of the electron is obtained from QED essentially through the fact that the Dirac matrix γ_μ of "semi-classical" theory (Chap. 1) is replaced by $\gamma_\mu + \Lambda_\mu$, where the convergent part of Λ_μ is $\Lambda_\mu^{(2)}$. The correction to the value of two for the Landé factor in Chap. 1 is made through a development of the term $\bar{u}(p')(\gamma_\mu + \Lambda_\mu^{(2)})u(p)$, where $\bar{u}(p')$ is an adjoint Dirac spinor (Chap. 1) and $u(p)$ is a spinor in the standard representation. The field $B^{(3)}$ emerges from the term $i\bar{u}(p')\sigma_{\mu\nu}q^\nu u(p)$, where $q := p' - p$ and where $\sigma_{\mu\nu}$ is defined by [16]

$$\sigma_{\mu\nu} = \frac{i}{2}(\gamma_\mu \gamma_\nu - \gamma_\nu \gamma_\mu). \tag{386}$$

The correction from renormalization, $\Lambda_\mu^{(2)}$, must be included in this term to calculate the field $B^{(3)}$ in QED, i.e., in a fully quantized theory of the interaction of one electron and one photon. Specifically, therefore, the photomagneton operator $\hat{B}^{(3)}$ emerges in QED through the interaction term

$$i\bar{u}(p')\frac{i}{2}(\gamma_\mu \gamma_\nu - \gamma_\nu \gamma_\mu)u(p), \tag{387}$$

where p' and p are momenta, the $u's$ denote spinors, and the $\gamma's$ denote Dirac matrices. The term $\Lambda_\mu^{(2)}$ reduces [16] to the dimensionless (S.I. units)

$$\Lambda_\mu^{(2)} = -\frac{e^2}{16\pi^2 \epsilon_0 \hbar c^2 m}(p_\mu + p_\mu'), \tag{388}$$

where ϵ_0 is the permittivity of free space and m is the electron mass. Therefore $\Lambda_\mu^{(2)}$ is a small correction to γ_μ obtained from renormalization and the removal of infinities. It involves the fine structure constant of spectroscopy,

$$\alpha = \frac{e^2}{4\pi\epsilon_0 \hbar c} = \frac{1}{137.0360}, \tag{389}$$

to first order, and the mass of the electron. The Landé factor of $g = 2$ from the original Dirac equation (Chap. 1) is corrected to

$$g = 2\left(1 + \frac{\alpha}{2\pi}\right), \tag{390}$$

to first order in α, the fine structure constant. From Eqs. (388) to (390), $\Lambda_\mu^{(2)}$ contains no reference to the electric and magnetic fields of electromagnetism, fields which are concomitant with the photon in QED, and therefore $\Lambda_\mu^{(2)}$ has no effect on $B^{(3)}$ and vice versa.

6.3.1 CALCULATION OF THE ANOMALOUS MAGNETIC MOMENT OF THE ELECTRON IN QED

The effect of the convergent vertex contribution $\Lambda_\mu^{(2)}$ is calculated in QED from the Dirac equation (Chap. 1),

$$\gamma_\mu \hat{p}_\mu u(p) = -mcu(p), \tag{391}$$

where $\hat{p}_\mu$ is the electron momentum operator. Equation (391) is written in S.I. units and in Minkowski notation, leading to a minus sign on the right hand side. It is an eigen-equation of quantum mechanics, the spinor $u(p)$ is an eigenfunction, and $\gamma_\mu \hat{p}_\mu$ an eigenoperator. Multiplying the equation on both sides by γ_ν produces

$$\gamma_\nu \gamma_\mu \hat{p}_\mu u(p) = -\gamma_\nu mcu(p), \tag{392}$$

which, with the definitions [16]

$$\begin{aligned} \gamma_\mu \gamma_\nu + \gamma_\nu \gamma_\mu &= 2g_{\mu\nu} := \{\gamma_\mu, \gamma_\nu\}, \\ \gamma_\mu \gamma_\nu - \gamma_\nu \gamma_\mu &= -2i\sigma_{\mu\nu} := [\gamma_\mu, \gamma_\nu], \end{aligned} \tag{393}$$

gives Eq. (391) in the form

$$\gamma_\nu u(p) = -\frac{1}{m_0 c}(\hat{p}_\nu - i\sigma_{\mu\nu}\hat{p}_\mu)u(p). \tag{394}$$

It is the term in $-i\sigma_{\mu\nu}\hat{p}_\mu$ that gives rise to the intrinsic electron spin (Chap. 1), essentially because $-i\bar{u}\sigma_{\mu\nu}\bar{u}$ is an antisymmetric (spin) tensor, and it is this term that is corrected by the factor $(1 + \frac{\alpha}{2\pi})$ in QED through consideration of the convergent vertex $\Lambda_\mu^{(2)}$. The correction takes place through

$$j_\mu = \overline{u}(p')\gamma_\mu u(p) \to \overline{u}(p')(\gamma_\mu + \Lambda_\mu^{(2)})u(p), \quad (395)$$

i.e., a conserved Dirac current (Chap. 1) describing an electron with Landé factor two is changed with precision to one with Landé factor of about 2.002. The calculation of this 1% change in the Landé factor takes place as follows in the Dirac equation (391) without the introduction of the electron-photon interaction energy, which contributes a term in the Lagrangian of the type

$$\mathcal{L}_{int} = -\frac{e}{\hbar} A_\mu \overline{\psi}\gamma_\mu \psi. \quad (396)$$

Since $B^{(3)}$ is contained within A_μ, there is no effect, inter alia, of this correction on $B^{(3)}$ and of $B^{(3)}$ on the magnitude of the correction, which is known from spectroscopic data with great precision.

Equation (394) is considered along with

$$\overline{u}(p')\gamma_\nu = -\frac{\overline{u}(p')}{m_0 c}(p'_\nu + i\sigma_{\mu\nu} p'_\mu), \quad (397)$$

in which p is defined through the Feynman parameter, Z [16],

$$p' = p - kZ. \quad (398)$$

These considerations lead to

$$j_\nu = -\frac{1}{2m_0 c}\left(\overline{u}(p')(p_\nu + p'_\nu)u(p) + i\overline{u}(p')\sigma_{\mu\nu}(p'_\nu - p_\nu)u(p)\right), \quad (399)$$

for the Dirac current before renormalization with $\Lambda_\mu^{(2)}$. The latter produces the renormalized current

$$j_\mu^{(R)} = \overline{u}(p')(\gamma_\mu + \Lambda_\mu^{(2)})u(p)$$
$$= -\frac{1}{2m_0 c}\overline{u}(p')\left(p_\mu + p'_\mu + i\left(1 + \frac{\alpha}{2\pi}\right)\sigma_{\mu\nu}(p'_\mu - p_\mu)\right)u(p), \quad (400)$$

and results in Eq. (390) for the Landé factor g. This calculation is carried out in the absence of interaction

$B^{(3)}$ and the Electron's Magnetic Moment

between electron and photon, and cannot therefore affect the field $B^{(3)}$. The latter makes its appearance as the expectation value of the photomagneton $\hat{B}^{(3)}$ through the Dirac equation (Chap. 1),

$$\gamma_\mu(p_\mu + eA_\mu)u(p) = -mcu(p), \tag{401}$$

for the interaction of an electron with an electromagnetic field represented by A_μ. Equation (401) leads before renormalization to the spin Hamiltonian

$$H_{spin} = 2\left(\frac{e\hbar}{4m}\right)\boldsymbol{\sigma} \cdot \langle\hat{B}^{(3)}\rangle, \tag{402}$$

where

$$B^{(3)} = \langle\hat{B}^{(3)}\rangle, \tag{403}$$

and where

$$\mu_e = \frac{e\hbar}{4m} \tag{404}$$

is the intrinsic magnetic dipole moment of one electron. The effect of introducing the convergent vertex $\Lambda_\mu^{(2)}$ into Eq. (401) is to change Eq. (402) to

$$H_{spin}^{(R)} = 2\left(1 + \frac{\alpha}{2\pi}\right)\left(\frac{e\hbar}{4m}\right)\boldsymbol{\sigma} \cdot B^{(3)}, \tag{405}$$

which leaves $B^{(3)}$ unchanged as expected. As discussed in Vol. 1, $\hat{B}^{(3)}$ is a constant of motion and commutes with the Hamiltonian; it is therefore also unaffected by light squeezing in quantum optics [5].

6.3.2 ORIGIN OF THE CONVERGENT VERTEX $\Lambda_\mu^{(2)}$ IN QED

The only particles in QED are photons and electrons, and $\Lambda_\mu^{(2)}$ emerges from one of the three primitive divergences [16] through use of dimensional normalization of the Lagrangian,

Chapter 6. $B^{(3)}$ in Quantum Electrodynamics

$$\mathcal{L} = i\bar{\psi}\gamma_\mu\partial_\mu\psi - mc^2\bar{\psi}\psi - \frac{e}{\hbar}A_\mu\bar{\psi}\gamma_\mu\psi$$

$$-\frac{\epsilon_0}{4}(\partial_\mu A_\nu - \partial_\nu A_\mu)^2 - \frac{1}{2}(\partial_\mu A_\mu)^2, \tag{406}$$

in which the Dirac spinor is represented by ψ and its adjoint by $\bar{\psi}$. Renormalization proceeds through the replacement,

$$e \rightarrow \mu^{2-d/2}e, \tag{407}$$

i.e., by changing only e, the electron charge. The $\Lambda_\mu^{(2)}$ factor then emerges from a vertex graph [16] of the type $-ie\Lambda_\mu(p, q, p+q)$. The removal of infinities of this graph results in a change of the physical properties of the electron, e.g. its mass and charge. The convergent $\Lambda_\mu^{(2)}$ is that part of the overall vertex with no k in the numerator of the integrand, and results in

$$\gamma_\mu^{(R)} = \gamma_\mu - \frac{e^2}{16\pi^2\epsilon_0\hbar mc^2}(p_\mu + p'_\mu), \tag{408}$$

i.e., in $\gamma_\mu \rightarrow \gamma_\mu^{(R)}$ after renormalization in QED. These considerations rigorously reinforce the conclusion in Chap. 1 that $B^{(3)}$ is a direct result of the Dirac equation describing the interaction of a quantized electron with the electromagnetic field, represented by A_μ. The rigorous and accurate methods of QED show that $\hat{B}^{(3)}$ is the quantized field property responsible for the formation of H_{spin} of Chap. 1 from the quantized electron spin whenever an electron interacts with a photon.

Chapter 7. Summary of Arguments and Suggestions for Experimental Verification

In these two volumes and eight hundred equations or so we have developed the theory of $\boldsymbol{B}^{(3)}$ and non-Abelian electrodynamics in the vacuum, using arguments drawn from several areas of contemporary field theory. Since $\boldsymbol{B}^{(3)}$ is a physical magnetic flux density in an axis (3) orthogonal to the plane of conventional vacuum electrodynamics, it indicates that the quantized field (the photon) carries mass. This is a consequence of special relativity, which asserts that a massless photon has only two degrees of physical polarization, and two helicities, 1 and -1. Conventional electrodynamics therefore does not self-consistently allow the existence of $\boldsymbol{B}^{(3)}$, although the latter emerges from the former through the conjugate product.

The conventional view that $\boldsymbol{B}^{(3)}$ is zero is contradicted by experience, namely in the phenomena of magnetization by circularly polarized electromagnetic radiation. This was illustrated in Chap. 1 of Vol. 1 using the inverse Faraday effect, with a theory based on the conjugate product $\boldsymbol{B}^{(1)} \times \boldsymbol{B}^{(2)}$, which is $iB^{(0)}\boldsymbol{B}^{(3)*}$. This phenomenon and others like it indicate experimentally, therefore, that $\boldsymbol{B}^{(3)}$ is real and non-zero, an inference which was reinforced in Chap. 12 of Vol. 1 and Chap. 1 of this volume using classical and quantum relativistic equations of motion of a single electron in the electromagnetic field, i.e., of e in A_μ. These rigorous calculations from first principles show that the spin and orbital angular momenta of the electron is governed *entirely* by $\boldsymbol{B}^{(3)}$, acting at first and second order in $B^{(0)}$. Furthermore, these calculations have defined with precision the experimental conditions under which the characteristic square root power density dependence (denoted $I_0^{1/2}$) of $\boldsymbol{B}^{(3)}$ dominates. Essentially, $I_0^{1/2}$ is observed in magnetization of an electron plasma by circularly polarized microwaves under the condition $\omega \leq (e/m)B^{(0)}$ where e/m is the charge to mass ratio of the electron, ω the angular frequency of the electromagnetic beam, and $B^{(0)}$ its magnetic flux density amplitude. This condition has already been approached quite closely by Deschamps et al. [43b] in their demonstration of the inverse Faraday effect, and by increasing the power density of the beam by a factor of about ten to a hundred, the $I_0^{1/2}$ dependence dominates theoretically in this experiment. In the first part of this summary chapter we discuss the precise conditions under which the $I_0^{1/2}$ dependence emerges in this important experiment. There is a clear need for such an experiment in order to prove the existence of $B^{(3)}$ unequivocally under precise conditions, possible with contemporary technology. The simple (but exact) one electron theories of these volumes can be refined if necessary to allow for statistical effects of many electrons in a plasma, but they are already adequate to describe the major results of such an experiment. The theoretically expected demonstration of the $I_0^{1/2}$ dependence of $B^{(3)}$ would be a major experimental advance in field theory, and the physics of fields and particles in general. Such a demonstration would render vacuum electrodynamics a non-Abelian, three dimensional, theory in the vacuum, and possibly to unification of electrodynamics with general relativity. Furthermore, it would indicate that the quantized electromagnetic field carries mass, and this would lead to further experimental support for spontaneous symmetry breaking of the vacuum.

The key experiment consists of magnetization of an electron plasma with a pulse of microwave radiation of peak power of about ten to one hundred megawatts. Furthermore, this experiment is possible in practice through a relatively simple adjustment of the conditions described by Deschamps et al. [43b], using thirty year old technology. These authors demonstrated magnetization of an electron plasma formed from an inert gas by a megawatt peak power pulse of 30 GHz radiation. A 100 turn induction coil detected the current due to magnetization stemming from the field $B^{(3)}$ as discussed in Chap. 12 of Vol. 1. The microsecond pulses were detected using a synchronized oscilloscope. The plasma was created in a pyrex tube of helium gas 0.065 m in diameter and 0.2 m long, linked into a circular wave guide carrying circularly polarized microwave radiation. The area of the sample was therefore about 0.003 square meters. For peak microwave power of a megawatt therefore, the peak power density was about 3×10^8 W m^{-2}, producing a peak $B^{(0)}$ of about 0.002 tesla (see Chap. 12 of Vol. 1). Using e/m for the electron of roughly 2×10^{11} C kgm^{-1} we obtain $eB^{(0)}/m$ to be about 4×10^8 rad s^{-1}. The microwave frequency of 30 GHz corresponds to an angular frequency of roughly 2×10^{10} rad s^{-1}. Therefore, under these typical reported [43b] conditions we obtain

$$\omega \sim 50 \frac{e}{m} B^{(0)}. \tag{409}$$

In consequence we expect the magnetization $\boldsymbol{M}^{(3)}$ from the Hamilton-Jacobi equation (Chap. 12 of Vol. 1) to be dominated by an I_0 dependence, and this is reported experimentally in Fig. (2) of Ref. [43b].

This observation is in itself enough to prove the existence of $\boldsymbol{B}^{(3)}$, through the equation

$$\boldsymbol{M}^{(3)} \sim N\beta'' B^{(0)} \boldsymbol{B}^{(3)}, \tag{410}$$

where β'' is a one electron hyperpolarizability, and where there are N non-interacting electrons in the plasma. Equation (410) is the $\omega \gg \frac{e}{m} B^{(0)}$ limit of the following result from the relativistic Hamilton-Jacobi equation of e in A_μ

$$\boldsymbol{M}^{(3)} = \frac{Ne^3 c^2}{2m\omega^2} \left(\frac{B^{(0)}}{(m^2\omega^2 + e^2 B^{(0)2})^{\frac{1}{2}}} \right) \boldsymbol{B}^{(3)}. \tag{411}$$

The characteristic $I_0^{1/2}$ dependence is obtained, however, in the limit

$$\omega \leq \frac{e}{m} B^{(0)}, \tag{412}$$

which gives

$$\boldsymbol{M}^{(3)} \sim N\chi' \boldsymbol{B}^{(3)}, \tag{413}$$

where χ' is a one electron susceptibility. It is possible to achieve condition (412) experimentally using the same basic apparatus as Deschamps *et al.*, but with an increased power density for the same frequency. The increased power density (or intensity) can be achieved by increasing the peak pulse power and by narrowing the sample area while maintaining the frequency at 30 GHz. According to Eq. (411) the observed magnetization will become a mixture of terms in $I_0^{1/2}$ and I_0 as condition (412) is approached, and will eventually become dominated by the $I_0^{1/2}$ dependence. This will leave no doubt as to the existence of the vacuum $\boldsymbol{B}^{(3)}$, because an $I_0^{1/2}$ dependence cannot be obtained from plane waves such as $\boldsymbol{B}^{(1)}$ and $\boldsymbol{B}^{(2)}$, which time average to zero at first order in $B^{(0)}$.

Therefore not only is this a critically important experiment (as is the optical Aharonov-Bohm experiment) but it is also relatively straightforward with contemporary technology.

Apart from the experiment of Deschamps et al. using microwaves, the presently available experimental indications of the existence of $\boldsymbol{B}^{(3)}$ are based on phenomena at visible frequencies. The simple but rigorous calculations developed in Chap. 12 of Vol. 1 and recounted above of the spin trajectory of e in A_μ show that at visible frequencies, the magnetization is produced always under the condition $\omega \gg (e/m)B^{(0)}$ and is therefore always dominated by the term in $B^{(0)}\boldsymbol{B}^{(3)}$ (Eq. (410)). This means that $\boldsymbol{M}^{(3)}$ in the visible range is proportional always to I_0 for all but the most intense laser pulses. This explains why the I_0 dependence dominates in the experimental data of van der Ziel et al. [43a] obtained some thirty years ago in liquids and solids. These series of experiments first demonstrated the inverse Faraday effect using focused, giant ruby laser pulses. These data, as is now realized, provide evidence for $B^{(0)}\boldsymbol{B}^{(3)}$ in the vacuum. Similar phenomena such as light shifts [50], and the optical Faraday and Zeeman effects [4,5,7] are dominated at visible frequencies by an I_0 dependence provided that a correctly relativistic treatment is developed of these phenomena. The conventional description of a phenomenon such as the inverse Faraday effect depends on a semi-classical approach [12] using the conjugate product $\boldsymbol{B}^{(1)} \times \boldsymbol{B}^{(2)}$ well known in nonlinear optics [5]. The key discovery [1–10],

$$\boldsymbol{B}^{(1)} \times \boldsymbol{B}^{(2)} = iB^{(0)}\boldsymbol{B}^{(3)*}, \qquad (414)$$

means that the conjugate product is equated with $iB^{(0)}\boldsymbol{B}^{(3)*}$. The existence of $\boldsymbol{B}^{(3)}$ is obscured, however, in the conventional semi-classical theory, which, furthermore, is given usually [12] in terms of

$$\boldsymbol{E}^{(1)} \times \boldsymbol{E}^{(2)} = c^2 \boldsymbol{B}^{(1)} \times \boldsymbol{B}^{(2)}, \qquad (415)$$

a cross product which seems at first sight to be remote from any magnetic field. This was first shown to be proportional to $B^{(3)}$ in 1992 [1,4]. A typical semi-classical description of the inverse Faraday effect is that of Woźniak et al. [12]. This is non-relativistic, and the term in $I_0^{1/2}$ is missing completely. It is clearly necessary to reappraise carefully the techniques of magneto-optics in order to make the theory rigorously relativistic. Only then will $B^{(3)}$ emerge through its $I_0^{1/2}$ dependence in a self-consistent way.

Some experimental features at visible frequencies were sketched out in Chap. 7 of Vol. 1 and discussed there in terms of $B^{(3)}$ in a non-relativistic framework. The proper relativistic approach is typified in the equations of motion of e in A_μ, the Hamilton-Jacobi and Dirac equations.

Atomic and molecular matter is thought to be composed of nuclei and electrons arranged according to the Pauli exclusion principle in orbitals, and a rigorous approach to the $I_0^{1/2}$ dependence in these systems requires a solution of the Dirac equation with the appropriate N electron Hamiltonian. The inverse Faraday effect has been evaluated in atomic systems by Kielich et al. [51] using a non-relativistic numerical method, but there is at present no work available on the rigorous solution of the Dirac equation (e.g. in its Hamilton-Jacobi form) for magnetization by microwave pulses of atomic matter. This is the next step up from plasma (free electrons) but will probably require the methods of computational physics applied to the Dirac equation rather than the Schrödinger equation, methods which have been extensively developed by, for example, Clementi et al. [62]. Similarly, there are few data available on magnetization by light, and none in the required condition (412) as discussed already, although Deschamps et al., thirty three years ago, came remarkably close.

It is overwhelmingly probable that the $I_0^{1/2}$ dependence from Eq. (413) will be observed experimentally because if not, the Hamilton-Jacobi equation itself will have failed. This outcome is vanishingly improbable because of the vast amount of data from other sources in favor of this classical equation of motion, first devised non-relativistically in the eighteen thirties. This line of thought traces the existence of $B^{(3)}$ to the principle of least action, upon which the relativistic version of the equation is based. This illustrates how deeply imbedded is $B^{(3)}$ in classical physics, provided it is approached in a suitably relativistic way. In relativistic quantum physics, it is likewise a direct result

Chapter 7. Summary and Suggestions

of the Dirac equation (Chap. 1) and is in consequence *as fundamental* as the intrinsic electron spin itself. This inference was rigorously reinforced in QED in the previous chapter.

Similarly, the existence of $\boldsymbol{B}^{(3)}$ in the vacuum leads to the expectation (Chap. 8 of Vol. 1) of an optical Aharonov-Bohm effect. This would detect the vector potential, $\boldsymbol{A}_3$, due to $\boldsymbol{B}^{(3)}$ after gauge transformation in to areas where $\boldsymbol{B}^{(3)}$ itself is excluded. Again, however, a precise treatment of this effect is relativistic, as in the original paper by Aharonov and Bohm, reviewed in Chap. 8 of Vol. 1. This precise treatment is required in order to optimize the chances of successfully detecting $\boldsymbol{B}^{(3)}$ in such an experiment. Overall, however, the OAB parallels the ordinary AB effect, now well-verified experimentally after forty-four years of exploration. To design a successful OAB experiment requires careful estimates and maximization of the sensitivity of the detection system. One method [63] that might succeed involves the modification of a SQUID device. The most obvious method, discussed in Vol. 1, involves passing a circularly polarized beam of radiation in the shadow of interfering electron beams. From our considerations above, this may well have to be at microwave frequencies, but at the time of writing the relativistic theory is not fully developed.

The magnetic properties of electromagnetic radiation are therefore summarized in the fact that its phase independent magnetic field, $\boldsymbol{B}^{(3)}$, does not average to zero at first order in $B^{(0)}$, the scalar magnitude of the magnetic flux density of the beam. Unlike $\boldsymbol{B}^{(1)}$ and $\boldsymbol{B}^{(2)}$, the ordinary plane waves, $\boldsymbol{B}^{(3)}$ does not time average to zero, even at the highest frequencies. This leads to its characteristic $I_0^{1/2}$ dependence when the beam interacts with matter, typified in the simplest case by one electron. The relativistic nature of this interaction means that the $I_0^{1/2}$ dependence can be seen clearly only under the condition (412). Therefore the same inference carries through to atomic and molecular matter, where e is bound in orbitals and not free. The fact that $\boldsymbol{B}^{(3)}$ does not time average to zero is responsible for optical NMR, discussed briefly in Chap. 7 of Vol. 1. Optical NMR is beginning to be understood, and has been observed experimentally [64], but with *visible* frequency lasers. If we consider the electron to be replaced by a nuclear particle such as a proton, 1800 times heavier with equal and opposite charge, and consider the interaction of e^+ in A_μ with the relativistic Hamilton-Jacobi equation of motion, the condition (412) will occur 1800 times lower in frequency for the same $B^{(0)}$. This thought experiment shows that no $I_0^{1/2}$ dependence will emerge from ONMR when a visible frequency laser is used. Under the conditions used for the first, exploratory, ONMR experiments [64] it is clear that any bulk shift will be dominated by an I_0 dependence, as in the inverse Faraday effect. Although such bulk shifts were reported [64], they are obscured by site specific effects, which are useful analytically, but which interfere with a demonstration of $B^{(3)}$ using this technique.

By comparison with the methods of Deschamps et al. [43b] it becomes clear that the $I_0^{1/2}$ dependence in a proton plasma or polarized proton beam will become dominant at radio frequencies for a peak pulse power density of about 10^8 or 10^9 W m^{-2}. It may be possible to explore these effects experimentally with contemporary technology, but the use of an electron plasma is technically much more straightforward, because microwave pulses can be used.

In previous work [4,5] one of us has initiated the study of $B^{(3)}$ in magneto-optics using the standard semi-classical approach [5,12]. This type of theory led to the theoretical prediction [4,5] of several novel effects, occurring with an $I_0^{1/2}$ dependence. Examples include the optical Faraday, Zeeman, Cotton-Mouton and Majorana effects, optical NMR and ESR, and the optical Aharonov-Bohm effect. In view of the relativistic effects just discussed, these semi-classical theories must be viewed as approximations, but ones which nevertheless lead to the correct $I_0^{1/2}$ dependence. The pioneering theory of Pershan [65], on the other hand, does not contain an $I_0^{1/2}$ dependence, and will not under any circumstances reduce to the result (411) of the correctly relativistic approach. Again, in the conventional semi-classical approach typified in Ref. [12], no $I_0^{1/2}$ dependence emerges.

The correctly relativistic description of the magnetizing properties of electromagnetic radiation, especially using microwave frequencies, must be based on the Dirac equation, and will probably be computationally intensive as discussed already. With contemporary computers this is not a problem. In the conventional semi-classical theories it is now clear that the conjugate product signals the existence of $iB^{(0)}B^{(3)*}$, and therefore of $B^{(3)}$, whose $I_0^{1/2}$ dependence, generated in its interaction with matter, must be calculated, however, relativistically. In this sense, $iB^{(0)}B^{(3)*}$ is observed whenever the conjugate product is observed, and since this is the antisymmetric part of light intensity itself [5], $iB^{(0)}\boldsymbol{B}^{(3)*}$ has been observed, with hindsight, on countless occasions, whenever circularly polarized radiation has been used.

This inference marks the end of the *Abelian era* in electrodynamics and electromagnetic field theory, a claim that can be tested experimentally as discussed already in this chapter through the existence of the $I_0^{1/2}$ dependence of $\boldsymbol{B}^{(3)}$. Clearly, if $\boldsymbol{B}^{(3)}$ is observed in this experiment, its presence in *all* magneto-optic effects [5] will have been signalled unequivocally and conclusively. There is hardly any need to emphasize further the importance and extensive consequences of such an outcome.

The existence, then, of $\boldsymbol{B}^{(3)}$ in the vacuum indicates that O(2) electrodynamics is internally inconsistent. In this way, O(2) gauge geometry, a flat geometry, *self indicates* that it is incomplete, and that the photon is a particle with three physical degrees of polarization in the vacuum. These are associated with the three physical magnetic fields $\boldsymbol{B}^{(1)}$, $\boldsymbol{B}^{(2)}$ and $\boldsymbol{B}^{(3)}$, the first two of which are plane waves, and the third of which is a spin field. This inference in turn *self indicates* that the photon must have mass, however tiny in magnitude, because a massless boson has only *two* degrees of polarization, in flat contradiction to our three dimensional world. As first shown by Wigner [22] a massless particle means an E(2) little group, *an entirely unphysical result*. The unequivocal experimental detection of the $I_0^{1/2}$ dependence using microwave pulses, or some other means, can therefore be taken to mean that the photon is massive, thus settling a debate that stretches over many scientific generations. This in turn might lead to renewed efforts in astronomy and relativistic cosmology to see the effect of photon mass, effects such as Tolman's tired light. Such efforts have been reviewed recently by Vigier [34]. The existence of $\boldsymbol{B}^{(3)}$ is in contradiction with the structure of Wigner's E(2) little group unless the boson (photon) being described by such a symmetry acquires mass.

Because of the powerful results of field theory [16] achieved in recent years, results based on gauge invariance, it is asserted by many field theorists that the photon mass (m_0) must be identically zero, and that light must have an infinite range. The reason is that the term $m_0 A_\mu A_\mu$ is not gauge invariant unless m_0 is identically zero. We differ from these colleagues in our inference of finite photon mass, because as argued, the emergence of an $I_0^{1/2}$ dependence *from* the *first principles* governing the trajectory of e in A_μ leads to the presence of **B**$^{(3)}$ in the vacuum. These dynamical principles are imbedded deeply in classical and quantum mechanics, and can be traced to the principle of least action, as shown in these volumes. These are powerful arguments in favor of a third dimension for electromagnetism in the vacuum, and therefore for fields concomitant with the photon. The inherently three dimensional Proca equation is the only one that can deal consistently with the emergence of **B**$^{(3)}$ and link it directly to photon mass through the Yukawa type exponential decay described in Chap. 2 of this volume. The d'Alembert equation on the other hand is one in a flat, two dimensional world, in which waves are purely transverse. The Proca equation emerges in Lagrangian theory from a vacuum whose symmetry is spontaneously broken in a way first described by Higgs and others. The breaking of the vacuum symmetry occurs in a Lagrangian which is originally *compatible* with gauge invariance. There is no reason therefore, to assert that photon mass must be zero in a symmetry broken vacuum. The origin of mass resides as usual in a Higgs mechanism in this breaking of vacuum symmetry, to which *all* fields are subject.

We have shown that **B**$^{(3)}$ = $B^{(0)}$ $exp(-\xi Z)$**k** is a solution of the Proca equation, where $\xi = m_0 c/\hbar$, thus relating **B**$^{(3)}$ and m_0, the photon mass. We have taken at face value the main result of the Higgs mechanism as applied to Abelian electromagnetism, that the photon acquires mass simultaneously with a third degree of polarization. The free photon is therefore a massive boson in the symmetry broken vacuum, for which there is firm evidence from particle physics as discussed in Chap. 5 of this volume. The vacuum itself is therefore a topologically non-trivial entity. Even if SSB is not used in electromagnetic theory, we have suggested that the condition $A_\mu A_\mu = 0$, a limiting form of the Dirac condition for vanishing photon radius, can be used to make finite photon mass compatible with gauge invariance.

Since SSB is so successful in GWS theory, however, it can be taken as having been proven experimentally in the CERN experiment [57] which detected $W_{\pm\mu}$ and Z_0 at their predicted masses. It is therefore logical to accept that SSB of a gauge theory leads to a finite photon mass in the vacuum, and that this indicates three degrees of polarization, of which **B**$^{(3)}$ is an experimentally provable sign. Reversing the argument, **B**$^{(3)}$ leads back to finite photon mass and a symmetry broken vacuum, and is, then, a manifestation of a third, physical dimension which has been long neglected. Clearly, we disagree with that part of GWS which models out the photon mass as identically zero, and encourage efforts such as those of Huang [35] to incorporate m_0 consistently in GWS and SU(5).

As argued in these volumes, the existence of the vacuum $\boldsymbol{B}^{(3)}$ introduces, ultimately, some profound philosophical modifications in our contemporary understanding of electrodynamics. Most obviously, $\boldsymbol{B}^{(3)}$ has no existence in O(2) gauge geometry, which therefore asserts that although $\boldsymbol{B}^{(1)} \times \boldsymbol{B}^{(2)}$, the conjugate product, is non-zero, $iB^{(0)}\boldsymbol{B}^{(3)*}$ is zero. This is a reduction to absurdity because $\boldsymbol{B}^{(1)} \times \boldsymbol{B}^{(2)}$ is equal, algebraically, to $iB^{(0)}\boldsymbol{B}^{(3)*}$. Absurdities of this nature result in O(2) electrodynamics because by definition its gauge geometry is planar. In other words, a physical vacuum field is not allowed perpendicular to this plane. Analytical algebra leads logically, however, to the central result of these volumes,

$$\boldsymbol{B}^{(1)} \times \boldsymbol{B}^{(2)} = iB^{(0)}\boldsymbol{B}^{(3)*}, \text{ and cyclic permutations,} \quad (416)$$

and these relations are inherently non-Abelian, indicating immediately the need for an O(3) gauge geometry for electromagnetism in the vacuum (Chap. 3 and 4 of this volume). The need for $m_0 = 0$ disappears in O(3) gauge geometry, and indeed, is a logical contradiction. Conservative proponents of O(2) gauge geometry cannot accept equation (416), nor are they allowed by their adherence to O(2) to accept that $\boldsymbol{B}^{(3)}$ emerges from first principles, i.e., from the equations (both classical and quantum mechanical) that describe the orbital and intrinsic spinning motion of e in A_μ. They are forced into an illogical corner, and must abandon the O(2) gauge geometry so long in favor as soon as the characteristic $I_0^{1/2}$ dependence of $\boldsymbol{B}^{(3)}$ is observed.

Secondly, the non-Abelian structure of Eqs. (416) has the major potential advantage of bringing vacuum electrodynamics into the same philosophical ball park as vacuum gravitation, described by general relativity (Appendix C). This is an interesting prospect which has been thought up to now to be on the distant horizon because of the inherently non-linear nature of general relativity.

Thirdly, there exists in O(3) gauge geometry the quantization condition $eA^{(0)} = \hbar\kappa$ in the vacuum. The O(2) photon momentum $\hbar\kappa$ becomes identified in O(3) with $eA^{(0)}$. The latter obviously has the units of linear momentum, but introduces the charge e, multiplying the vector potential amplitude $A^{(0)}$. The conventional view asserts that the electromagnetic gauge field is uncharged, and is the agent of interaction between two electrons. In QED the same process is described as an exchange of virtual photons, which are also uncharged. The O(3) relation $eA^{(0)} = \hbar\kappa$ on the other hand divides the quantum mechanical $\hbar\kappa$ into the product of two $\hat{C}$ negative quantities. This means that as in all non-Abelian field theories, the field is its own source, which in this case is the current caused by the *field charge e* moving through the vacuum. More generally, e is g, the $\hat{C}$ negative coupling constant of O(3) gauge geometry [16] and $g/\hbar$ in Eq. (260) of this volume is identified with $A^{(0)}/\kappa$. In this sense, the O(3) scaling constant simply becomes the classical quantity $A^{(0)}/\kappa$, and the charge quantization condition becomes $g = \hbar(\kappa/A^{(0)})$. Thus, g occurs in units of $\hbar$, the Dirac constant. Analogously, energy and linear momentum in O(2) theory also occur in units of $\hbar$, this being the Planck law of 1900. In O(3), the momentum $eA^{(0)}$ propagates through the vacuum with the photon. In this sense, Jackson [44] has shown, in his classical textbook, that an electronic charge moving infinitesimally close to the speed of light produces transverse plane waves which are entirely indistinguishable from those concomitant with the photon. Even within the O(2) framework used by Jackson, the field is its own source, an electron moving essentially at the speed of light. If we apply the minimal prescription that electromagnetic four-momentum, p_μ, becomes $p_\mu + eA_\mu$ in the presence of an electron, and consider the electron itself to be moving infinitesimally close to the speed of light, the result is $p_\mu = eA_\mu$, i.e., the four-momentum of electron and field become indistinguishable, because the field concomitant with the photon and the field generated by an electron travelling near c are indistinguishable. Applying quantization finally to $\boldsymbol{p}$ we obtain $\hbar\kappa = eA^{(0)}$, which is precisely the result from O(3) gauge geometry. The essential difference between the O(3) theory and Jackson's is of course that $\boldsymbol{B}^{(3)}$ is present self-consistently in O(3) gauge geometry, whereas it is unconsidered by Jackson. In the non-Abelian view of vacuum electromagnetism, the field carries its own source, just as gravitation self-propagates in general relativity [16]. This view is based on the fact that $\boldsymbol{A}^{(1)} \times \boldsymbol{A}^{(2)}$ is classically non-zero, and is directly proportional to the vacuum $\boldsymbol{B}^{(3)}$. Two electrons repel through the vacuum in the conventional view, even though they may be separated by very large distances. Classically, one electron becomes the source of a Coulomb force on the other and vice versa, action and reaction being equal and opposite. The force is transmitted by a force field, the electromagnetic field, and the latter can be detected only by the mutual repulsion felt by the two electrons. If the two electrons were, for the sake of argument, uncharged, there would be no repulsion and no field. Therefore the Coulomb force field depends on the existence of charge on the electron, which in the conventional quantum field theory propels a photon at the speed of light through a vacuum. The opposite process also occurs and there is an exchange of virtual photons. The photon is conventionally massless, with momentum $\hbar\kappa$ and intrinsic angular momentum $\hbar$. Equation (416) and the arguments proposed in these volumes now demand that the photon momentum $\hbar\kappa$ be identified with $eA^{(0)}$, and this relation quantizes charge. It is a direct outcome of using O(3) gauge geometry for vacuum electrodynamics. The source of vacuum electromagnetism, infinitely distant in O(2) gauge geometry, becomes identified with photon momentum itself in O(3) gauge geometry, and travels with the photon. The momentum of the photon becomes indistinguishable from $eA^{(0)}$, and in a manner of speaking, from its own source. A similar situation occurs in the self-propagating gravitational field in general relativity, as described by the Einstein field equations.

We hope that these ideas will be accepted in the spirit of free enquiry and that they will, accordingly, be tested experimentally. In particular, there is an urgent need to search for the characteristic $I_0^{1/2}$ dependence generated when $B^{(3)}$ from microwave pulses interacts with an electron plasma. Nature shows!

Appendix A. The O(3) Electromagnetic Field Tensor, $G_{\mu\nu}$, in the Circular Basis (1), (2), (3)

From fundamental gauge field theory (Chap. 3), the covariant derivative of the n component field ψ is given by

$$\frac{D\psi}{dx_\mu} = \partial_\mu \psi - i\frac{e}{\hbar} \hat{J}^a A_\mu^a \psi, \tag{A1}$$

as in Eq. (262) of the text. In the O(3) group

$$(\hat{J}^a)_{mn} = -i\epsilon_{amn}, \tag{A2}$$

which gives the (mn)'th element of the a'th rotation generator of the O(3) group. In O(3) the n component field ψ is a three component field denoted by the vector $\boldsymbol{\phi}$, the m'th component of which is [16], from Eqs. (A1) and (A2)

$$D_\mu \phi_m = \partial_\mu \phi_m - \frac{e}{\hbar}\epsilon_{amn}A_\mu^a \phi_n. \tag{A3}$$

This is the m'th component of a vector equation in the (isospin) space of $\boldsymbol{\phi}$,

$$D_\mu \boldsymbol{\phi} = \left(\partial_\mu + \frac{e}{\hbar}\mathbf{A}_\mu \times \right)\boldsymbol{\phi}. \tag{A4}$$

In order to derive (A4) from Eq. (A3), the following has been used

$$(\mathbf{A}_\mu \times \boldsymbol{\phi})_m = -\epsilon_{amn}A_\mu^a \phi_n. \tag{A5}$$

On the left hand side, the m'th component of the vector product $\mathbf{A}_\mu \times \boldsymbol{\phi}$ is equated to the tensor product on the right hand side, where, as usual, summation over repeated indices (a and n) is used. Equation (A5) is therefore an equation in vector components in a three dimensional isospin space in which $\mathbf{A}_\mu$ and $\boldsymbol{\phi}$ obey the ordinary vector algebra of three dimensional space. The symbol $\mathbf{A}_\mu$ therefore denotes a vector in this isospin space, carrying a *dummy index* μ which plays no direct or specific part in the vector algebra of the isospin space. The quantity A_μ^a in tensor notation is analogous with a connection coefficient in general relativity [16]. The Levi-Civita symbol ϵ_{amn} is as usual zero when any two subscripts are identical.

In this Appendix the isospin vector $\boldsymbol{\phi}$ is identified with A_μ *itself*, and the isospin vector algebra developed in the circular basis (1), (2), (3), rather than the Cartesian basis X, Y, Z appropriate to Eq. (A4). Finally, the isospin space in the basis (1), (2), (3) is identified with the configuration space in the same basis. This procedure is justified by the fact that the J^a matrices in the general equation (A1) become rotation generators of the group O(3) in a three dimensional *configuration space* of the laboratory frame of reference. In a Cartesian basis, the O(3) field tensor is

$$G_{\mu\nu} = \partial_\mu \mathbf{A}_\nu - \partial_\nu \mathbf{A}_\mu + \frac{e}{\hbar} \mathbf{A}_\mu \times \mathbf{A}_\nu, \qquad (A6)$$

which is an equation in the vectors $G_{\mu\nu}$, $\mathbf{A}_\nu$, and $\mathbf{A}_\mu$ of the isospin space. The vector cross product appearing on the right hand side can be expressed in tensor notation by

$$(\mathbf{A}_\mu \times \mathbf{A}_\nu)_a = \epsilon_{abc} A_\mu^b A_\nu^c, \qquad (A7)$$

where the a'th component of the cross product on the left hand side is generated on the right hand side by a tensor product in the isospin space, with summation over the repeated indices b and c. The indices μ and ν on both sides of Eq. (A7) refer to a different four dimensional space-time, are unaffected by vector or tensor manipulations in the isospin space. Thus, in Eq. (A6), $G_{\mu\nu}$ is a three component vector in the three dimensional isospin space. The indices $\mu\nu$ indicate that it is also a four-component tensor in four dimensional space-time.

In the *circular* basis (1), (2) and (3) the vector cross product is the *conjugate product* described at length in Vol. 1 and preceding chapters,

The O(3) Electromagnetic Field Tensor $G_{\mu\nu}$

$$\mathbf{A}^{(1)} \times \mathbf{A}^{(2)} = iA^{(0)2}\mathbf{e}^{(3)*} = iA^{(0)}\mathbf{A}^{(3)*}, \tag{A8}$$

where the characteristic factor i has appeared on the right hand side. This is due to the fact that unit vectors in (1), (2) and (3) obey the cyclic relations

$$\mathbf{e}^{(1)} \times \mathbf{e}^{(2)} = i\mathbf{e}^{(3)*}, \quad \mathbf{e}^{(3)*} = -i\mathbf{e}^{(1)} \times \mathbf{e}^{(2)}, \text{ et cyclicum}, \tag{A9}$$

where * denotes "complex conjugate". As described in Vol. 1, this basis is natural for the description of circular polarization in electromagnetic radiation. Unit vectors in the Cartesian representation, on the other hand, obey

$$\mathbf{i} \times \mathbf{j} = \mathbf{k}, \text{ et cyclicum}, \tag{A10}$$

in which there is no factor i on the right hand side. The circular and Cartesian bases are linked by

$$\mathbf{e}^{(1)} = \mathbf{e}^{(2)*} = \frac{1}{\sqrt{2}}(\mathbf{1} - i\mathbf{j}), \quad \mathbf{e}^{(3)*} = \mathbf{e}^{(3)} = \mathbf{k}, \tag{A11}$$

and by

$$\mathbf{i} \cdot \mathbf{i} + \mathbf{j} \cdot \mathbf{j} + \mathbf{k} \cdot \mathbf{k} = \mathbf{e}^{(1)} \cdot \mathbf{e}^{(1)*} + \mathbf{e}^{(2)} \cdot \mathbf{e}^{(2)*} + \mathbf{e}^{(3)} \cdot \mathbf{e}^{(3)*}, \tag{A12}$$

and are therefore equivalent. However, the transition,

$$\mathbf{e}^{(1)} e^{i\phi} \rightarrow \mathbf{e}^{(2)} e^{i\phi}, \tag{A13}$$

(where ϕ is the electromagnetic phase) switches the sense of circular polarization, and therefore the circular unit vectors $\mathbf{e}^{(1)}$, $\mathbf{e}^{(2)}$ and $\mathbf{e}^{(3)}$ are natural representations of the spatial distributions of the field. In the circular representation of isospin space, therefore,

$$\left(\mathbf{A}_\mu^{(1)} \times \mathbf{A}_\nu^{(2)}\right)_Z = \epsilon_{Zbc} A_{\mu b}^{(1)} A_{\nu c}^{(2)} = iA^{(0)} \mathbf{A}_{\mu\nu}^{(3)*}, \tag{A14}$$

and Eq. (A6) becomes

$$G_{\mu\nu}^{(3)*} = F_{\mu\nu}^{(3)*} - i\frac{e}{\hbar}\mathbf{A}_\mu^{(1)} \times \mathbf{A}_\nu^{(2)}, \tag{A15}$$

which is Eq. (276c). As in the text, Eq. (A15) reduces finally to

$$B^{(3)*} = B^{(3)} = -i\frac{e}{\hbar}A^{(1)} \times A^{(2)},$$ (A16)

which, except for an arbitrary sign change, is also Eq. (147) from the Dirac equation of Chap. 1.

We arrive at the important conclusion that the structure of Eq. (A16) links the O(2) and O(3) theories of electrodynamics, in that $A^{(1)}$ and $A^{(2)}$ exist in the O(2) (Abelian) theory, and therefore so does the conjugate product $A^{(1)} \times A^{(2)}$. However, this same O(2) conjugate product gives $B^{(3)}$, the spin field, adding a third physical dimension in isospin space, identified with configuration space. *This third dimension can be accommodated self-consistently only within the structure of O(3) (non-Abelian) gauge theory.* The existence of $B^{(3)}$ is *indicated*, however, by the existence of $B^{(1)}$ and $B^{(2)}$, through the cyclic relations of Vol. 1,

$$B^{(1)} \times B^{(2)} = iB^{(0)}B^{(3)*}, \text{ et cyclicum,}$$ (A17)

so that O(2) theory "self-indicates" that it is incomplete. The experimental basis for the physical nature of $B^{(1)} \times B^{(2)}$ is the inverse Faraday effect, whose rigorous derivation from first principles was demonstrated classically in Chap. 12 of Vol. 1, and quantum mechanically in Chap. 1 of this volume.

It is therefore of the utmost importance to note that the inverse Faraday effect, and other magnetic effects of light indicate *experimentally* that electrodynamics is non-Abelian in nature. The expected experimental observation of the characteristic square root power density *fingerprint* of $B^{(3)}$ also signals the transition from O(2) to O(3) electrodynamics, with many consequences throughout field theory.

Appendix B. The O(3) Covariant Derivative (D_μ) in the Basis (1), (2), (3)

In NAE the O(2) derivative operator ∂_μ becomes D_μ, the O(3) covariant derivative operator, and it is basically important to define D_μ self-consistently in any isospin basis, including the circular basis (1), (2), (3). From fundamental gauge theory [16], D_μ is defined in general by Eq. (262) of the text, which in O(3) reduces in the Cartesian basis to

$$D_\mu \psi_m = \partial_\mu \psi_m - i\frac{e}{\hbar}(\hat{J}^a)_{mn}, \tag{B1}$$

for the m'th component of the field ψ. The latter can be identified with the three component vector A_μ in isospin space, a vector whose m'th scalar component is $A_{\mu m}$. In Eq. (B1), $(\hat{J}^a)_{mn}$ denotes the mn'th element of the a'th Cartesian rotation generator of O(3). These infinitesimal generators are defined for a = 1, 2, and 3 by (Vol. 1),

$$\hat{J}^1 = \begin{bmatrix} 0 & 0 & 0 \\ 0 & 0 & -i \\ 0 & i & 0 \end{bmatrix}, \quad \hat{J}^2 = \begin{bmatrix} 0 & 0 & i \\ 0 & 0 & 0 \\ -i & 0 & 0 \end{bmatrix},$$

$$\hat{J}^3 = \begin{bmatrix} 0 & -i & 0 \\ i & 0 & 0 \\ 0 & 0 & 0 \end{bmatrix}. \tag{B2}$$

With these definitions, Eq. (B1) reduces in the Cartesian basis to

$$D_\mu A_{\nu m} = \partial_\mu A_{\nu m} - i\frac{e}{\hbar}(\hat{J}^a)_{mn} A_{\mu a} A_{\nu n}. \tag{B3}$$

In the *circular* basis, however, the conjugate product is defined by

$$(\mathbf{A}^{(1)} \times \mathbf{A}^{(2)})_{(3)} = i\hat{J}_{mn}^3 A_n^{(1)} A_m^{(2)}. \tag{B4}$$

In this equation, the left hand side denotes the (3)'th scalar component of the vector conjugate product. On the right hand side, $\hat{J}^3$ is defined by Eq. (B2), $A_n^{(1)}$ is the n'th Cartesian component of $\mathbf{A}^{(1)}$, $A_m^{(2)}$ is the m'th Cartesian component of $\mathbf{A}^{(2)}$. Thus

$$(\mathbf{A}^{(1)} \times \mathbf{A}^{(2)})_{(3)} = A_X^{(1)} A_Y^{(2)} - A_Y^{(1)} A_X^{(2)}, \tag{B5}$$

and

$$i\hat{J}_{nm}^3 = \epsilon_{3nm}. \tag{B6}$$

Therefore

$$\begin{aligned} D_\mu A_{\nu m} &= \partial_\mu A_{\nu m} - i\frac{e}{\hbar} \hat{J}_{nm}^c A_{\mu n}^{(a)} A_{\nu m}^{(b)} \\ &= \partial_\mu A_{\nu m} - \frac{e}{\hbar} \left(\mathbf{A}_\mu^{(a)} \times \mathbf{A}_\nu^{(b)}\right)_{(c)}. \end{aligned} \tag{B7}$$

In the circular basis, a vector cross product generates an *imaginary* quantity through

$$-i\mathbf{e}^{(1)} \times \mathbf{e}^{(2)} = \mathbf{e}^{(3)*}, \text{ et cyclicum,} \tag{B8}$$

and therefore the vector form of Eq. (B7) is

$$\begin{aligned} D_\mu \mathbf{A}_\nu^{(3)*} &= \partial_\mu \mathbf{A}_\nu^{(3)*} - i\frac{e}{\hbar} \mathbf{A}_\mu^{(1)} \times \mathbf{A}_\nu^{(2)}, \\ D_\mu \mathbf{A}_\nu^{(2)*} &= \partial_\mu \mathbf{A}_\nu^{(2)*} - i\frac{e}{\hbar} \mathbf{A}_\mu^{(3)} \times \mathbf{A}_\nu^{(1)}, \\ D_\mu \mathbf{A}_\nu^{(1)*} &= \partial_\mu \mathbf{A}_\nu^{(1)*} - i\frac{e}{\hbar} \mathbf{A}_\mu^{(2)} \times \mathbf{A}_\nu^{(3)}. \end{aligned} \tag{B9}$$

The vector cross product can be expressed by

$$\mathbf{A}_\mu^{(a)} \times \mathbf{A}_\nu^{(b)} = iA_\mu \mathbf{A}_\nu^{(c)*}, \tag{B10}$$

so that Eq. (B7) becomes

The O(3) Covariant Derivative (D_μ)

$$D_\mu A_\nu^{(c)*} = \partial_\mu A_\nu^{(c)*} + \frac{e}{\hbar} A_\mu A_\nu^{(c)*}, \tag{B11}$$

for c = 1, 2 and 3. Therefore, in the circular basis,

$$D_\mu := \partial_\mu + \frac{e}{\hbar} A_\mu, \tag{B12}$$

where A_μ is a scalar in the isospin space, but a four-vector in space-time. Note that the equivalent result in the Cartesian basis [16] is

$$(D_\mu)_{Cartesian} = \partial_\mu - i\frac{e}{\hbar} A_\mu. \tag{B13}$$

In the *circular* basis, the field tensor and D operator components are linked, furthermore, by

$$G_{\mu\nu} = \frac{\hbar}{e}[D_\mu, D_\nu] = \partial_\mu A_\nu - \partial_\nu A_\mu + \frac{e}{\hbar}[A_\mu, A_\nu], \tag{B14}$$

so that

$$G_{\mu\nu}^{(a)} = \partial_\mu A_\nu^{(a)} - \partial_\nu A_\mu^{(a)} - i\frac{e}{\hbar}\left[A_\mu^{(b)}, A_\nu^{(c)}\right], \tag{B15}$$

a result which is obtained from the unit vector component relations

$$e_Z^{(3)*} = -i\left(e_X^{(1)} e_Y^{(2)} - e_X^{(2)} e_Y^{(1)}\right)_Z, \tag{B16}$$

and cyclic permutations of (1), (2) and (3). Equation (B15) in vector notation for the isospin space becomes

$$G_{\mu\nu}^{(1)*} = \partial_\mu \mathbf{A}_\nu^{(1)*} - \partial_\nu \mathbf{A}_\mu^{(1)*} - i\frac{e}{\hbar} \mathbf{A}_\mu^{(2)} \times \mathbf{A}_\nu^{(3)},$$

$$G_{\mu\nu}^{(2)*} = \partial_\mu \mathbf{A}_\nu^{(2)*} - \partial_\nu \mathbf{A}_\mu^{(2)*} - i\frac{e}{\hbar} \mathbf{A}_\mu^{(3)} \times \mathbf{A}_\nu^{(1)}, \tag{B17}$$

$$G_{\mu\nu}^{(3)*} = \partial_\mu \mathbf{A}_\nu^{(3)*} - \partial_\nu \mathbf{A}_\mu^{(3)*} - i\frac{e}{\hbar} \mathbf{A}_\mu^{(1)} \times \mathbf{A}_\nu^{(2)},$$

which are Eqs. (276) of the text. *Equations (B17) do not reduce to the conventional O(2) definitions of the electromagnetic field tensor, because the conjugate products are always non-zero.*

Appendix C. The Structural Analogy between NAE and General Relativity

C.1 THE COVARIANT DERIVATIVES

The covariant derivatives used in Appendices A and B and Chap. 3 and following are modelled [16] on general relativity. For curved space-time, the axes themselves vary from point to point, and for a contravariant vector V^μ, its covariant derivative is in general relativity

$$D_\nu V^\mu = \partial_\nu V^\mu + \Gamma^\mu_{\lambda\nu} V^\lambda, \tag{C1}$$

where $\Gamma^\mu_{\lambda\nu}$ are analogous to A^a_μ of NAE [16]. If O(3) is extended to the Lorentz group in NAE, the structure of A^a_μ becomes the same as that of the connection coefficients of general relativity. The NAE isospin space becomes four dimensional space-time, and not the three dimensional configurational space (1), (2) and (3) of the text. The covariant derivative of NAE then becomes the same in structure as its counterpart in general relativity. In both NAE and general relativity, connection coefficients would then connect the components of a vector at one point with its components at a nearby point, the vector being transported between points by parallel transport [16]. Such properties could be useful in the search for a unified and self consistent understanding of gravitation and electromagnetism. Electrodynamical laws could then be developed as laws of general relativity and vice-versa.

C.2 THE CURVATURE TENSOR

The analogue of the non-Abelian $G^a_{\mu\nu}$ is the Riemann-Christoffel curvature tensor defined by

$$R^{\kappa}_{\lambda\mu\nu} = \partial_\nu \Gamma^{\kappa}_{\lambda\mu} - \partial_\mu \Gamma^{\kappa}_{\lambda\nu} + \Gamma^{\rho}_{\lambda\mu}\Gamma^{\kappa}_{\rho\nu} - \Gamma^{\rho}_{\lambda\nu}\Gamma^{\kappa}_{\rho\mu},$$ (C2)

which is to be compared with

$$G^{a}_{\mu\nu} = \partial_\mu A^{a}_\nu - \partial_\nu A^{a}_\mu + e\epsilon_{abc} A^{b}_\mu A^{c}_\nu.$$ (C3)

The curvature tensor indicates that space-time becomes curved in the presence of gravitation. Analogously, we would in a unified understanding be able to say that space-time becomes curved in the presence of electromagnetism.

C.3 THE BIANCHI IDENTITY

The Bianchi identity in general relativity is the analogue of the O(3) homogeneous Maxwell equations of Chap. 4, and they are, respectively

$$D_\rho R^{\kappa}_{\lambda\mu\nu} + D_\mu R^{\kappa}_{\lambda\nu\rho} + D_\nu R^{\kappa}_{\lambda\rho\mu} = 0,$$ (C4)

and

$$D_\rho G_{\mu\nu} + D_\mu G_{\nu\rho} + D_\nu G_{\rho\mu} = 0,$$ (C5)

which in O(2) electrodynamics becomes [16]

$$\partial_\rho F_{\mu\nu} + \partial_\mu F_{\nu\rho} + \partial_\nu F_{\rho\mu} = 0,$$ (C6)

the vacuum homogeneous Maxwell equations of the conventional O(2) theory.

C.4 THE EINSTEIN FIELD EQUATIONS

The curvature of space-time in general relativity is determined by the canonical energy-momentum tensor $T^{\mu\nu}$, which appears in the Einstein field equations,

$$R_{\mu\nu} = \frac{1}{2} g_{\mu\nu} R - \frac{8\pi G}{c^2} T_{\mu\nu},$$ (C7)

where $R_{\mu\nu}$ is the Ricci tensor, $g_{\mu\nu}$ is the metric tensor, and

NAE and General Relativity

G the gravitation constant. In O(3) electrodynamics A_μ becomes proportional in the vacuum to the energy-momentum vector p_μ through the charge quantization condition of Chaps. 3 and 4, and there is a link between p_μ and $T_{\mu\nu}$, both being energy momentum tensors. In O(2) electrodynamics, eA_μ is a momentum vector p_μ only in the presence of matter, in the simplest case an electron with charge e. The O(3) equations are naturally non-linear, through the cyclic relations

$$\boldsymbol{B}^{(1)} \times \boldsymbol{B}^{(2)} = iB^{(0)}\boldsymbol{B}^{(3)*}, \text{ et cyclicum,} \qquad (C8)$$

and similar as developed throughout these two volumes. They are naturally analogous with the non-linear general theory of relativity. Roughly speaking, gravitation and light are driven through the vacuum in the same way in these non-linear theories, both fields being self-generating.

Appendix D. Structure of the Field Tensor $G_{\mu\nu}^{(i)}$ of Non-Abelian Electrodynamics

In this Appendix, it is shown that the $G_{\mu\nu}^{(i)}$ tensor of non-Abelian electrodynamics reduces to the $F_{\mu\nu}^{(i)}$ tensor of Abelian electrodynamics using the charge quantization condition

$$e\mathbf{A}^{(i)} = \hbar\partial^{(i)}. \tag{D1}$$

Vacuum solutions of Maxwell's equations are plane waves, which in the $F_{\mu\nu}$ tensor of conventional Abelian electrodynamics are asserted to be purely transverse to the direction of propagation, and described by the U(1) symmetry group. Vacuum solutions of the non-Abelian Maxwell equations (Chap. 4) include these plane waves, and, self-consistently, the vacuum field $\mathbf{B}^{(3)}$ and its unphysical dual $-i\mathbf{E}^{(3)}/c$. The overall result of this Appendix is that in the vacuum

$$G_{\mu\nu}^{(i)} = 2F_{\mu\nu}^{(i)}, \quad (i) = (1), (2), \tag{D2}$$

for transverse circular states of the electromagnetic radiation. In Abelian electrodynamics, the $F_{\mu\nu}^{(3)}$ tensor is conventionally a null tensor, but its non-Abelian generalization, $G_{\mu\nu}^{(3)}$, contains $\mathbf{B}^{(3)}$ and $-i\mathbf{E}^{(3)}/c$ as elements.

These results are obtained self-consistently from the O(3) vector potential through the equation

$$G_{\mu\nu}^{(1)*} = \partial_\mu^{(0)} \times \mathbf{A}_\nu^{(1)*} - i\partial_\mu^{(j)} \times \mathbf{A}_\nu^{(k)}, \tag{D3}$$

which splits into three equations for scalar components as follows

147

$$G_{\mu\nu}^{(1)*} = \left[\partial_\mu^{(0)}, A_\nu^{(1)*}\right] - i\left[\partial_\mu^{(2)}, A_\nu^{(3)}\right],$$

$$G_{\mu\nu}^{(2)*} = \left[\partial_\mu^{(0)}, A_\nu^{(2)*}\right] - i\left[\partial_\mu^{(3)}, A_\nu^{(1)}\right], \qquad (D4)$$

$$G_{\mu\nu}^{(3)*} = \left[\partial_\mu^{(0)}, A_\nu^{(3)*}\right] - i\left[\partial_\mu^{(1)}, A_\nu^{(2)}\right],$$

(using the notation developed in Chap. 5). These equations show that $G_{\mu\nu}^{(1)*}$ is a sum of two four-curls. In this notation, the conventional $F_{\mu\nu}$ tensor of U(1) electrodynamics is

$$F_{\mu\nu}^{(1)*} = \left[\partial_\mu^{(0)}, A_\nu^{(1)*}\right], \qquad F_{\mu\nu}^{(2)*} = \left[\partial_\mu^{(0)}, A_\nu^{(2)*}\right],$$

$$F_{\mu\nu}^{(3)*} = \left[\partial_\mu^{(0)}, A_\nu^{(3)*}\right] = 0, \qquad (D5)$$

for each state of circular polarization. The general form of $F_{\mu\nu}$ can be displayed conventionally as the four by four matrix

$$F_{\mu\nu} = \begin{bmatrix} 0 & B_Z & -B_Y & -i\dfrac{E_X}{c} \\ -B_Z & 0 & B_X & -i\dfrac{E_Y}{c} \\ B_Y & -B_X & 0 & -i\dfrac{E_Z}{c} \\ i\dfrac{E_X}{c} & i\dfrac{E_Y}{c} & i\dfrac{E_Z}{c} & 0 \end{bmatrix}. \qquad (D6)$$

D.1 CIRCULAR STATE (3): MAGNETIC FIELDS

D.1.1 XY AND YX COMPONENTS

These components of $G_{\mu\nu}^{(3)*}$ contain the field $\boldsymbol{B}^{(3)*}$ through

$$B_Z^{(3)*} = B_Z^{(3)} = \left[\partial_X^{(0)}, A_Y^{(3)*}\right] - i\left[\partial_X^{(1)}, A_Y^{(2)}\right] = G_{XY}^{(3)*} = -G_{YX}^{(3)*}, \qquad (D7)$$

which reduces to

$$B_Z^{(3)} = G_{XY}^{(3)*} = -i(\partial^{(1)} \times \boldsymbol{A}^{(2)})_Z := (\boldsymbol{\nabla}^{(1)} \times \boldsymbol{A}^{(2)})_Z, \qquad (D8)$$

because $\boldsymbol{A}^{(3)}$ only has a Z component by definition. Without

Field Tensor $G^{(1)}_{\mu\nu}$ of Non-Abelian Electrodynamics

the second commutator of O(3) theory therefore, $\boldsymbol{B}^{(3)}$ is zero. Moreover, the second commutator reduces to the cyclic relations between three physical magnetic fields through which the presence of $\boldsymbol{B}^{(3)}$ was first discovered [1–9],

$$\boldsymbol{B}^{(3)} = \boldsymbol{B}^{(3)*} = -i\frac{e}{\hbar}\boldsymbol{A}^{(1)} \times \boldsymbol{A}^{(2)} = -i\frac{\kappa}{A^{(0)}}\boldsymbol{A}^{(1)} \times \boldsymbol{A}^{(2)}$$

$$= -i\partial^{(1)} \times \boldsymbol{A}^{(2)} = -\frac{i}{B^{(0)}}\boldsymbol{B}^{(1)} \times \boldsymbol{B}^{(2)}.$$

(D9)

D.1.2 XZ AND ZX COMPONENTS

These are defined through

$$G^{(3)*}_{XZ} = -G^{(3)*}_{ZX} = \left[\partial^{(0)}_X, A^{(3)*}_Z\right] - i\left[\partial^{(1)}_X, A^{(2)}_Z\right] = 0,$$

(D10)

and vanish because $\boldsymbol{A}^{(3)}$ is divergentless and irrotational, and because $\boldsymbol{A}^{(2)}$ has only X and Y components through its definition as a transverse plane wave

$$\boldsymbol{A}^{(2)} = \boldsymbol{A}^{(1)*} = \frac{A^{(0)}}{\sqrt{2}}(-i\boldsymbol{1} + \boldsymbol{j})e^{-i\phi}, \qquad \phi := \omega t - \boldsymbol{\kappa} \cdot \boldsymbol{r}.$$

(D11)

D.1.3 YZ AND ZY COMPONENTS

Similarly, these components vanish, so that the only non-zero magnetic component is $\boldsymbol{B}^{(3)}$.

D.2 CIRCULAR STATE (3): ELECTRIC FIELDS

D.2.1 Z4 AND 4Z COMPONENTS

In Abelian electrodynamics, these would formally contain a real electric field $\boldsymbol{E}^{(3)}_Z$ multiplied by i, whose presence is due to the Minkowski algebra of special relativity. The elements are given in non-Abelian (O(3)) electrodynamics by

$$G_{Z4}^{(3)*} = \left[\partial_Z^{(0)}, A_4^{(3)*}\right] - i\left[\partial_Z^{(1)}, A_4^{(2)}\right],$$

$$G_{4Z}^{(3)*} = \left[\partial_4^{(0)}, A_Z^{(3)*}\right] - i\left[\partial_4^{(1)}, A_Z^{(2)}\right].$$

(D12)

The first (conventional Abelian) commutator in these equations vanishes because $\mathbf{A}^{(3)}$ is divergentless and irrotational, leaving

$$G_{Z4}^{(3)*} = \partial_Z^{(1)} A_4^{(2)} - \partial_4^{(1)} A_Z^{(2)} = 0.$$

(D13)

This defines a putative real electric field, which as argued in the text, is zero on the grounds of elementary symmetry. Because $\mathbf{A}^{(2)}$ and $\partial^{(1)}$ have no Z components, the real electric field from the Z4 element vanishes self-consistently. Similarly, the real electric field from the 4Z element vanishes because $\mathbf{A}^{(2)}$ and $\partial^{(1)}$ have no Z components. Therefore the real part of the electric field in the (3) state is zero. An imaginary and unphysical electric field $-i\mathbf{E}^{(3)}/c$ is obtained formally, however, through the fact that it is dual to the real $\mathbf{B}^{(3)}$ as described in detail in Chap. 4.

D.2.2 X4 AND 4X ELEMENTS

The X component of $\mathbf{E}^{(3)}$ is clearly zero because $\mathbf{E}^{(3)}$ is in the Z direction, and similarly for $E_Y^{(3)}$. The general definition of $G_{\mu\nu}^{(3)*}$, (Eq. (D4)) must produce this result self-consistently. We have

$$G_{X4}^{(3)*} = -i\frac{E_X^{(3)*}}{c} = -i\left[\partial_X^{(1)}, A_4^{(2)}\right]$$

$$= -i\frac{e}{\hbar}\left(A_X^{(1)} A_4^{(2)} - A_4^{(1)} A_X^{(2)}\right) = 0.$$

(D14)

However, by definition of $\mathbf{A}^{(1)}$ and $\mathbf{A}^{(2)}$ as plane waves, we have

$$A_X^{(1)} = i\frac{A^{(0)}}{\sqrt{2}}e^{i\phi}, \qquad A_X^{(2)} = -i\frac{A^{(0)}}{\sqrt{2}}e^{-i\phi},$$

(D15)

and this indicates that

Field Tensor $G_{\mu\nu}^{(1)}$ of Non-Abelian Electrodynamics

$$A_4^{(1)} = A_4^{(2)} = 0, \tag{D16}$$

i.e., that the time-like components of the four-vectors defined by

$$A_\mu^{(1)} := \left(\mathbf{A}^{(1)}, iA_4^{(1)}\right), \quad A_\mu^{(2)} := \left(\mathbf{A}^{(2)}, iA_4^{(2)}\right), \tag{D17}$$

are zero. This result is consistent with the arguments developed for dual four-vectors in Chap. 11 of Vol. 1.

D.2.3 Y4 AND 4Y COMPONENTS

Similarly, the general definition (D4) gives

$$G_{Y4}^{(3)*} = -i\frac{E_Y^{(3)*}}{c} = -i\left[\partial_Y^{(1)}, A_4^{(2)}\right], \tag{D18}$$

and since $E_Y^{(3)*}$ must be zero if $\mathbf{E}^{(3)*}$ is directed in the Z axis, we obtain

$$A_Y^{(1)} A_4^{(2)} - A_4^{(1)} A_Y^{(2)} = 0. \tag{D19}$$

Comparing equations (D14) and (D19) shows that they are self-consistent if and only if Eq. (D16) holds, i.e., if the time-like components of $A_\mu^{(1)}$ and $A_\mu^{(2)}$ are rigorously zero. The same result follows for the well-known transverse gauge [16] in conventional O(2) electrodynamics, where $A_\mu^{(3)}$ is not considered.

The overall result for the (3) state therefore is

$$G_{\mu\nu}^{(3)*} = \begin{bmatrix} 0 & B_Z^{(3)*} & 0 & 0 \\ -B_Z^{(3)*} & 0 & 0 & 0 \\ 0 & 0 & 0 & -i\dfrac{E_Z^{(3)*}}{c} \\ 0 & 0 & i\dfrac{E_Z^{(3)*}}{c} & 0 \end{bmatrix}, \tag{D20}$$

where it is very important to note that the electric fields appearing in the matrix are *unphysical*, i.e., are pure imaginary and appear because they are formally dual to the real and *physical* $B^{(3)*}$. The physical electric fields from the definition (D4) vanish as argued already. The only non-zero magnetic component from Eq. (D4) is $B^{(3)*}$.

Therefore these results, obtained from the *rigorous* theory of O(3) gauge geometry, are precisely consistent with our arguments in Vol. 1, and in the first few chapters of this volume. The results show that the choice of (1), (2) and (3) as isospin indices is physically meaningful and self-consistent. The analysis shows that the time-like components of the four-vectors $A_\mu^{(1)}$ and $A_\mu^{(2)}$ are zero.

D.3 POLARIZATION STATE (1): MAGNETIC FIELDS

D.3.1 XY AND YX COMPONENTS

These are given by the elements

$$G_{XY}^{(1)*} = -G_{YX}^{(1)*} = [\partial_X^{(0)}, A_Y^{(1)*}] - i[\partial_X^{(2)}, A_Y^{(3)}] = 0. \qquad (D21)$$

The second commutator vanishes because $A^{(3)}$ has no X or Y components, and the first commutator vanishes because the transverse plane wave $A^{(1)}$ has no specific X and Y dependence. Therefore the XY and YX components of polarization state (1) vanish. Note that the elements of the $G_{\mu\nu}^{(1)}$ tensor reduce to those of the $F_{\mu\nu}^{(i)}$ tensor.

D.3.2 XZ AND ZX ELEMENTS

In this case the general definition (D4) produces

$$G_{XZ}^{(1)*} = -G_{ZX}^{(1)*} = [\partial_X^{(0)}, A_Z^{(1)*}] - i[\partial_X^{(2)}, A_Z^{(3)}], \qquad (D22)$$

which can be expressed as

$$G_{XZ}^{(1)*} = -B_Y^{(1)*} - i\frac{e}{\hbar}\left(A_X^{(2)} A_Z^{(3)} - A_Z^{(2)} A_X^{(3)}\right) = -B_Y^{(1)*}$$
$$- i\frac{e}{\hbar} A_X^{(2)} A_Z^{(3)} = F_{XZ}^{(1)*} - i\frac{e}{\hbar} A_X^{(2)} A_Z^{(3)}. \qquad (D23)$$

Field Tensor $G_{\mu\nu}^{(1)}$ of Non-Abelian Electrodynamics

Using the relations

$$A_Z^{(3)} = A^{(0)} = \frac{B^{(0)}}{\kappa}, \qquad (D24)$$

$$\frac{e}{\hbar} = \frac{\kappa}{A^{(0)}}, \qquad (D25)$$

Eq. (D23) reduces to

$$G_{XZ}^{(1)*} = -B_Y^{(1)*} - iB_X^{(2)}. \qquad (D26)$$

Using the definition of the magnetic fields $\boldsymbol{B}^{(1)}$ and $\boldsymbol{B}^{(2)}$ as plane waves,

$$\boldsymbol{B}^{(1)} = \boldsymbol{B}^{(2)*} = \frac{B^{(0)}}{\sqrt{2}}(i\boldsymbol{1} + \boldsymbol{j})e^{i\phi}, \qquad (D27)$$

we obtain the results

$$B_X^{(1)} = i\frac{B^{(0)}}{\sqrt{2}}e^{i\phi}, \qquad B_X^{(2)} = -i\frac{B^{(0)}}{\sqrt{2}}e^{-i\phi},$$

$$B_Y^{(1)} = \frac{B^{(0)}}{\sqrt{2}}e^{i\phi}, \qquad B_Y^{(2)} = \frac{B^{(0)}}{\sqrt{2}}e^{-i\phi}, \qquad (D28)$$

which show that

$$B_Y^{(1)*} = iB_X^{(2)} = B_Y^{(2)}, \qquad G_{XZ}^{(1)*} = F_{XZ}^{(1)*} + F_{XZ}^{(1)*}, \qquad (D29)$$

i.e., the $G_{XZ}^{(1)*}$ element reduces to a sum of identical $F_{XZ}^{(1)*}$ elements. Each commutator of Eq. (D22) gives the same $F_{XZ}^{(1)*}$ element.

Similarly, we obtain for the ZX element,

$$G_{ZX}^{(1)*} = F_{ZX}^{(1)*} + F_{ZX}^{(1)*}. \qquad (D30)$$

D.3.3 YZ AND ZY ELEMENTS

In a similar way, it can be shown that

$$G_{YZ}^{(1)*} = -G_{ZY}^{(1)*} = B_X^{(1)*} - i\kappa A_Y^{(2)} = B_X^{(1)*} - iB_Y^{(2)} \qquad \text{(D31)}$$
$$= B_X^{(2)} + B_X^{(2)} = F_{YZ}^{(1)*} + F_{YZ}^{(1)*},$$

and the $G_{YZ}^{(1)*}$ and $G_{ZY}^{(1)*}$ elements reduce to a sum of identical $F_{YZ}^{(1)*}$ and $F_{ZY}^{(1)*}$ elements respectively.

D.4 POLARIZATION STATE (1): ELECTRIC FIELDS

D.4.1 X4 AND 4X ELEMENTS

In this case

$$G_{X4}^{(1)*} = \left[\partial_X^{(0)}, A_4^{(1)*}\right] - i\left[\partial_X^{(2)}, A_4^{(3)}\right] = -i\frac{E_X^{(1)*}}{c} \qquad \text{(D32)}$$
$$- i\frac{e}{\hbar}A_X^{(2)}A_4^{(3)},$$

which reduces to

$$G_{X4}^{(1)*} = -i\frac{E_X^{(1)*}}{c} + B_X^{(2)} = F_{X4}^{(1)*} + F_{X4}^{(1)*}, \qquad \text{(D33)}$$

if and only if

$$A_4^{(3)} = iA^{(0)}, \qquad \text{(D34)}$$

i.e., if and only if the time-like component of the longitudinal four-vector $A_\mu^{(3)} := \left(\mathbf{A}^{(3)}, iA_4^{(3)}\right)$ is pure imaginary. These results are of course obtained with the definitions of $\mathbf{E}^{(1)}$ and $\mathbf{E}^{(2)}$ as conjugate plane waves (see texts of Vols. 1 and 2),

Field Tensor $G^{(1)}_{\mu\nu}$ of Non-Abelian Electrodynamics

$$E^{(1)} = E^{(2)*} = \frac{E^{(0)}}{\sqrt{2}}(\mathbf{1} - i\mathbf{j})e^{i\phi}. \tag{D35}$$

The condition (D34) emerges self-consistently, furthermore, from an analysis of the Y4 and 4Y elements as follows. We find that in each and every case, the elements of the $G^{(1)*}_{\mu\nu}$ tensor reduce to sums over identical $F^{(1)*}_{\mu\nu}$ tensor elements as for the magnetic fields in circular state (1). This is the overall result summarized in Eq. (291) of Chap. 4. We obtain the additional information that the time-like components of $A^{(1)}_{\mu}$ and $A^{(2)}_{\mu}$ are zero, and that the time-like component of $A^{(3)}_{\mu}$ is pure imaginary if all elements of $G^{(1)*}_{\mu\nu}$ in polarization state (1) are to reduce self-consistently to $F^{(1)*}_{\mu\nu}$ elements in the way described.

D.4.2 4Y AND Y4 ELEMENTS

From the general formula (D4) these elements are given by

$$G^{(1)*}_{Y4} = -G^{(1)*}_{4Y} = \left[\partial^{(0)}_Y, A^{(1)*}_4\right] - i\left[\partial^{(2)}_Y, A^{(3)}_4\right], \tag{D36}$$

which reduces to

$$G^{(1)*}_{Y4} = -i\frac{E^{(1)*}_Y}{c} - i\frac{e}{\hbar}A^{(2)}_Y A^{(3)}_4 = -i\frac{E^{(2)}_Y}{c} + B^{(2)}_Y \tag{D37}$$

$$= F^{(1)*}_{Y4} + F^{(1)*}_{Y4},$$

if the time-like component of the four-vector $A^{(3)}_{\mu}$ is pure imaginary, as in Eq. (D34). This result is obtained with the plane wave elemental relations

$$-i\frac{E^{(2)}_Y}{c} = -i\frac{E^{(1)*}_Y}{c} = B^{(2)}_Y. \tag{D38}$$

Therefore we again obtain

$$G^{(1)*}_{\mu\nu} = 2F^{(1)*}_{\mu\nu}, \quad \mu = Y, \quad \nu = 4, \tag{D39}$$

D.4.3 Z4 AND 4Z ELEMENTS

In this case

$$G_{Z4}^{(1)*} = \left[\partial_Z^{(0)}, A_4^{(1)*}\right] - i\left[\partial_Z^{(2)}, A_4^{(3)}\right], \qquad (D40)$$

and the first commutator vanishes because the 4 components of both $\mathbf{A}^{(1)*}$ and its Z component are both zero. This is the result obtained in conventional O(2) electrodynamics, because by definition, the transverse plane wave $\mathbf{E}^{(1)*}$ of polarization state (1) has no Z component. The second commutator vanishes because

$$\partial_Z^{(2)} = \frac{e}{\hbar} A_Z^{(2)} = 0, \qquad \partial_4^{(2)} = \frac{e}{\hbar} A_4^{(2)} = 0, \qquad (D41)$$

and yet again we obtain the result

$$G_{Z4}^{(1)*} = F_{Z4}^{(1)*} + F_{Z4}^{(1)*} = 0 \qquad (D42)$$

so that the G tensor element reduces to a sum of two identical F tensor elements. In this case, these elements are zero.

Therefore, for all elements of $G_{\mu\nu}^{(1)*}$, the general result is obtained that

$$G_{\mu\nu}^{(1)*} = 2F_{\mu\nu}^{(1)*} \text{ for all } \mu \text{ and } \nu \qquad (D43)$$

provided that

$$A_\mu^{(1)} = (\mathbf{A}^{(1)}, 0), \qquad A_\mu^{(2)} = (\mathbf{A}^{(2)}, 0),$$
$$A_\mu^{(3)} = (i\mathbf{A}^{(3)}, i(iA^{(0)})). \qquad (D44)$$

Field Tensor $G_{\mu\nu}^{(1)}$ of Non-Abelian Electrodynamics

D.5 POLARIZATION STATE (2)

Similarly, it can be shown that

$$G_{\mu\nu}^{(2)*} = 2F_{\mu\nu}^{(2)*} \tag{D45}$$

for all elements of state (2).

These results show that if it is assumed that $\mathbf{A}^{(1)}$ and $\mathbf{A}^{(2)}$ are plane waves of the type (D11) then the $G_{\mu\nu}^{(1)}$ tensor of O(3) gauge geometry reduces, for polarization states (1) and (2), to twice the $F_{\mu\nu}^{(i)}$ tensor of O(2) gauge geometry. Thus, as discussed in Chap. 4, the $\mathbf{B}^{(1)}$, $\mathbf{B}^{(2)}$, $\mathbf{E}^{(1)}$ and $\mathbf{E}^{(2)}$ fields are recovered unchanged, i.e., are the same in O(3) and O(2) theory.

For polarization state (3), however, which is rigorously inconsistent with O(2) or planar gauge geometry, $F_{\mu\nu}^{(3)}$ is conventionally zero or undefined, depending on the viewpoint. Equation (D9), however, shows that $\mathbf{B}^{(3)}$ is recovered self consistently from our general equation (D4), which is derived from the rigorous theory of gauge geometry. Thus, our general equation (D4) produces $\mathbf{B}^{(3)}$ and its unphysical dual $-i\mathbf{E}^{(3)}/c$ self consistently with the transverse fields in the vacuum. This result is accompanied by the further insight

$$A_4^{(1)} = A_4^{(2)} = 0, \quad A_4^{(3)} = iA^{(0)}, \tag{D46}$$

which show that $A_\mu^{(1)}$ and $A_\mu^{(2)}$ are polar four-vectors, while $A_\mu^{(3)}$ is an axial four-vector, or pseudo four-vector, of the type considered in detail in Chap. 11 of Vol. 1. This result is also rigorously consistent with the cyclic relations for vector potentials obtained in Chap. 1 of Vol. 1

$$\mathbf{A}^{(1)} \times \mathbf{A}^{(2)} = -A^{(0)}(i\mathbf{A}^{(3)})^*, \text{ et cyclicum.} \tag{D47}$$

Thus, the vector cross product of two polar vectors, $\mathbf{A}^{(1)}$ and $\mathbf{A}^{(2)}$, must produce an axial vector, which is the pure imaginary, $(i\mathbf{A}^{(3)})*$. The four-vector equivalent to this space-like component is therefore the axial four-vector with imaginary components

$$A_\mu^{(3)} = (i\mathbf{A}^{(3)}, i(iA^{(0)})) = i(\mathbf{A}^{(3)}, iA^{(0)}). \tag{D48}$$

Therefore the magnitudes $|i\mathbf{A}^{(3)}|$ and $|iA^{(0)}|$ are equal. This is precisely analogous with the well known Gupta-Bleuler condition as discussed in Vol. 1. The axial four-vector $A_\mu^{(3)}$ is therefore light-like. Since it has no real 4 component, the real scalar potential can be viewed as zero, and this is the conventional assumption of O(2) electrodynamics in either the transverse (Coulomb) gauge, or in the Lorentz gauge. However, in conventional O(2) theory, the imaginary $iA^{(0)}$ is also asserted to be zero, or left unconsidered.

Appendix E. Some Details of the Non-Abelian Maxwell Equations in the Vacuum

In this Appendix some structural details are given of the O(3) Maxwell equations in the vacuum, and a self-consistency check developed for the inhomogeneous equations of Chap. 4

$$D_\nu G^{(i)*}_{\mu\nu} = 0, \quad (i) = (1), (2), (3). \tag{E1}$$

We have seen in Chap. 4 and Appendix D that this can be written as

$$D_\nu F^{(i)*}_{\mu\nu} = 0, \tag{E2}$$

and since $\boldsymbol{B}^{(3)}$ is formed from the O(2) fields $\boldsymbol{B}^{(1)}$ and $\boldsymbol{B}^{(2)}$, Eq. (E2) must reduce in the transverse polarizations (1) and (2) to

$$\partial_\nu F^{(i)*}_{\mu\nu} = 0, \quad (i) = (1), (2). \tag{E3}$$

Since

$$D_\nu = \partial_\nu + \frac{e}{\hbar}A_\nu, \tag{E4}$$

and $eA_\nu \neq 0$, the transition from Eq. (E2) to (E3) is possible if and only if

$$\partial_\nu = \frac{e}{\hbar}A_\nu, \tag{E5}$$

which is the scalar form of the charge quantization condition used in Chaps. 3 and following, and in Appendix D.
More precisely, Eq. (E5) must be written as

$$\partial_\nu^{(i)} = \frac{e}{\hbar} A_\nu^{(i)}, \qquad (i) = (1), (2), (3), \tag{E6}$$

as in Appendix D. From Eqs., (E2) and (E6) we obtain

$$\partial_\nu^{(i)*} F_{\mu\nu}^{(i)*} = \frac{e}{\hbar} A_\nu^{(i)*} F_{\mu\nu}^{(i)*} = 0. \tag{E7}$$

Equation (E7) allows a check for self-consistency to be developed through the expectation that each μ component must vanish independently of

$$A_\nu^{(i)*} F_{\mu\nu}^{(i)*} = 0. \tag{E8}$$

When this check is complete in detail (as follows) the O(3) inhomogeneous Maxwell equations (E1) can be expressed simply as

$$\partial_\nu^{(i)*} F_{\mu\nu}^{(i)*} = 0. \tag{E9}$$

For (i) = (1) and (2) these are identical with the O(2) equations. For (i) = (3) they give the inhomogeneous part of the Maxwell equations for $\boldsymbol{B}^{(3)}$ and $-i\boldsymbol{E}^{(3)}/c$ as discussed in Chap. 4.

E.1 DETAILED CHECKS ON EQUATION (E8)

E.1.1 POLARIZATION STATE (3)

Eq. (E8) becomes

$$A_\nu^{(3)} F_{\mu\nu}^{(3)} = A_X^{(3)} F_{\mu X}^{(3)} + A_Y^{(3)} F_{\mu Y}^{(3)} + A_Z^{(3)} F_{\mu Z}^{(3)} + A_4^{(3)} F_{\mu 4}^{(3)}. \tag{E10}$$

For $\mu = X$ Eq. (E10) becomes (see Appendix D)

$$A_Z^{(3)} F_{XZ}^{(3)} + A_4^{(3)} F_{X4}^{(3)} = 0, \tag{E11}$$

and similarly for $\mu = Y, Z,$ and 4.

Some Details of Non-Abelian Maxwell Equations 161

E.1.2 POLARIZATION STATE (1)

For $\mu = X$ and $\mu = Y$ all elements vanish independently of

$$A_\nu^{(1)} F_{\mu\nu}^{(1)} = A_X^{(1)} F_{\mu X}^{(1)} + A_Y^{(1)} F_{\mu Y}^{(1)} + A_Z^{(1)} F_{\mu Z}^{(1)} + A_4^{(1)} F_{\mu 4}^{(1)}, \quad (E12)$$

for reasons given in detail in Appendix D. For $\mu = Z$

$$A_X^{(1)} F_{ZX}^{(1)} + A_Y^{(1)} F_{ZY}^{(1)} = A_X^{(1)} B_Y^{(1)} - A_Y^{(1)} B_X^{(1)}$$

$$= \frac{\left(B_X^{(1)} B_Y^{(1)} - B_Y^{(1)} B_X^{(1)}\right)}{\kappa} = 0. \quad (E13)$$

For $\mu = 4$,

$$A_X^{(1)} F_{4X}^{(1)} + A_Y^{(1)} F_{4Y}^{(1)} = \frac{i}{\kappa c}\left(B_X^{(1)} E_X^{(1)} + B_Y^{(1)} E_Y^{(1)}\right) = 0 \quad (E14)$$

because

$$B_X^{(1)} = i\frac{B^{(0)}}{\sqrt{2}} e^{i\phi}, \quad B_Y^{(1)} = \frac{B^{(0)}}{\sqrt{2}} e^{i\phi},$$

$$E_X^{(1)} = \frac{E^{(0)}}{\sqrt{2}} e^{i\phi}, \quad E_Y^{(1)} = -i\frac{E^{(0)}}{\sqrt{2}} e^{i\phi}. \quad (E15)$$

Similarly, it can be shown that Eq. (E8) is true for polarization state (2). This completes the check for self-consistency in the reduction of Eq. (E1) to (E9). These results are utilized in Chap. 4 and following parts of Vol. 2.

References

[1] M. W. Evans, *Physica B*, **182**, 227 (1992).
[2] M. W. Evans, *Physica B*, **182**, 237 (1992).
[3] M. W. Evans, *Physica B*, **183**, 103 (1993).
[4] M. W. Evans, *The Photon's Magnetic Field* (World Scientific, Singapore, 1992).
[5] M. W. Evans, and S. Kielich, eds., *Modern Nonlinear Optics*, Vol. 85(2) of *Advances in Chemical Physics*, I. Prigogine and S. A. Rice, eds. (Wiley Interscience, New York, 1993).
[6] M. W. Evans and A. A. Hasanein, *The Photomagneton in Quantum Field Theory*. Vol. 1 (World Scientific, Singapore, 1994).
[7] M. W. Evans and J.-P. Vigier, *The Enigmatic Photon Volume 1: The Field* $B^{(3)}$ (Kluwer, Dordrecht, 1994).
[8] M. W. Evans in A. van der Merwe and A. Garuccio, eds., *Waves and Particles in Light and Matter* (Plenum, New York, 1994).
[9] M. W. Evans, *Found. Phys. Lett.* **7**, 67 (1994); ibid., in press; ibid., in press; ibid., in press.
[10] M. W. Evans, *Found. Phys.* in press; ibid., in press; ibid., in press
[11] M. W. Evans, *Mod. Phys. Lett.* **7**, 1247 (1993).
[12] S. Woźniak, M. W. Evans, and G. Wagnière, *Mol. Phys.* **75**, 81, 99 (1992).
[13] P. W. Atkins, *Molecular Quantum Mechanics* 2nd edn. (Oxford University Press, Oxford, 1983).
[14] E. U. Condon and H. Odabasi, *Atomic Structure* (Cambridge University Press, Cambridge, 1980).
[15] W. G. Richards and P. R. Scott, *Structure and Spectra of Atoms* (Wiley, London, 1976).
[16] L. H. Ryder, *Quantum Field Theory* 2nd edn. (Cambridge University Press, Cambridge, 1987).
[17] A. O. Barut *Electrodynamics and Classical Theory of Fields and Particles* (McMillan, New York, 1964).
[18] see Atkins, Ref. 13, pp. 388 ff.
[19] B. L. Silver, *Irreducible Tensorial Methods* (Academic, New York, 1976).

[20] J. D. Bjorken and S. D. Drell, *Relativistic Quantum Mechanics* (McGraw-Hill, New York, 1964).
[21] L. D. Barron *Physica B* **190**, 307 (1993); replied to by M. W. Evans, ibid., 310.
[22] E. P. Wigner, *Ann. Math.* **40**, 149 (1939).
[23] P. A. M. Dirac, *Proc. Roy. Soc.* **A133**, 60 (1931).
[24] G. Herzberg, *Atomic Spectra and Atomic Structure.* (Dover, New York, 1944).
[25] E. P. Wigner, *Group Theory* (Academic, New York, 1959).
[26] A. S. Wightman in C. de Witt and M. Jacob, eds. *High Energy Physics* (Gordon and Breach, London, 1965).
[27] e.g. M. Aguilar-Benitez et al. *Rev. Mod. Phys.* **56**, 2, part II (1984).
[28] A. S. Goldhaber and M. M. Nieto, ibid., **43**, 277 (1971).
[29] P. W. Higgs, *Phys. Lett.* **12**, 132 (1964).
[30] D. Bohm and B. J. Hiley, *Il Nuovo Cim.* **52A**, 295 (1979).
[31] S. L. Glashow, *Nuc. Phys.* **22**, 579 (1961); S. Weinberg, *Phys. Rev. Lett.* **19**, 1264 (1967); A. Salam, in N. Svartholm ed. *Proceedings, 8th Nobel Symposium* (Almqvist and Wiksell, Stockholm, 1968).
[32] for further details see Refs. 5 and 8.
[33] Reviewed in Ref. 28.
[34] J.-P. Vigier, "Present experimental status of the Einstein-de Broglie theory of light," *Proceedings, 4th International Symposium on Foundations of Quantum Mechanics* M. Tsukada et al., eds. (Japanese Journal of Applied Physics, Tokyo, 1993).
[35] J. C. Huang, *J. Phys. G, Nucl. Phys.* **13**, 273 (1987).
[36] G. 't Hooft, *Nucl. Phys.* **B79**, 276 (1974); A. M. Polyakov, *JETP Lett.* **20**, 197 (1975).
[37] D. E. Soper, *Classical Field Theory* (Wiley, New York, 1976).
[38] M. Moles and J.-P. Vigier, *Comptes Rendues* **276**, 697 (1973).
[39] P. W. Higgs, *Phys. Rev. Lett.* **13**, 508 (1964).
[40] P. W. Higgs, *Phys. Rev.* **145**, 1156 (1966).
[41] Reviewed in Ref. 34.
[42] P. G. de Gennes, *Superconductivity of Metals and Alloys* (Benjamin, New York, 1966).
[43] H. O. van der Ziel, P. S. Pershan and L. D. Malmstrom *Phys. Rev. Lett.* **15**, 190 (1965); J. Deschamps, M. Fitaire, and M. Lagoutte, ibid., **25**, 1330 (1970) and *Rev. Appl. Phys.* **7**, 155 (1972).
[44] J. D. Jackson, *Classical Electrodynamics* (Wiley, New York, 1962).
[45] L. D. Landau and E. M. Lifshitz *The Classical Theory of Fields* 4th edn. (Pergamon, Oxford, 1976).

References

[46] M. Born and E. Wolf, *Principles of Optics* 6th edn. (Pergamon, Oxford, 1975).
[47] L. H. Ryder, *Elementary Particles and Symmetries* (Gordon and Breach, London, 1986).
[48] H. B. Nielsen and P. Olesen, *Nucl. Phys.* **B61**, 45 (1973).
[49] B. S. Mathur, H. Tang, and W. Happer, *Phys. Rev.* **171**, 11 (1968).
[50] reviewed by W. Happer, *Rev. Mod. Phys.* **44**, 169 (1972).
[51] reviewed by R. Zawodny, in Vol. 85(1) of Ref. 5.
[52] G. 't Hooft, *Phys. Rev. Lett.* **37**, 8 (1976).
[53] A. M. Polyakov, *Soviet Phys. JETP* **41**, 988 (1976).
[54] W. K. H. Panofsky and M. Phillips, *Classical Electricity and Magnetism* 2nd. edn. (Addison-Wesley, Reading, Mass. 1962).
[55] J. P. van der Ziel, P. S. Pershan, and L. D. Malmstrom, *Phys. Rev.* **143**, 574 (1966); and other papers discussed in Ref. 7.
[56] see Refs. 17 and 54.
[57] UA1 Collaboration, *Phys. Lett.* **122B**, 103; **126B**, 398 (1983).
[58] C. S. Wu, E. Ambler, R. Hayward, D. Hoppes, and R. Hudson, *Phys. Rev.* **105**, 1413 (1957).
[59] S. Weinberg, *Phys. Rev. Lett.* **19**, 1264 (1967).
[60] S. Weinberg, *Rev. Mod. Phys.* **52**, 515 (1980); A. Salam, ibid., 526; S. L. Glashow, ibid., 539.
[61] T. W. B. Kibble, *Phys. Rev.* **155**, 1554 (1967).
[62] E. Clementi ed., *MOTECC Series of volumes and software* (Escom, Leiden, 1988 to present).
[63] M. A. Novikov, Institute of Applied Physics, Russian Academy of Sciences, cummunication.
[64] W. S. Warren, S. Mayr, D. Goswami and A. P. West, Jr., *Science* **255**, 1683 (1992); **259**, 836 (1993); D. Goswami, Ph.D. Thesis, (Princeton University, 1994).
[65] P. S. Pershan, *Phys. Rev.* **130**, 919 (1963).

Index

't Hooft Polyakov monopoles 84

Abelian electromagnetism 131
Abrikosov line(s) 41, 48
Abstract isospin space (1, 2, 3) 105
Aharonov-Bohm effect 68
Ampère's equation 54
Angular momentum operators 15
Anomalous
  magnetic moment of the electron 113, 118
  Zeeman effect 1
Anti-particles 1, 25
Antisymmetric part of light intensity 130
Astronomy and relativistic cosmology 130

$B^{(3)}$
  a rotation generator of the Poincaré group 52
  a topological string 48
  a topological vortex in free space 43
  an indicator of photon mass 110
  and canonical quantization 113
  and characteristic $I_0^{1/2}$ dependence 128, 130, 132, 134
  and its square root power density dependence 39
  and Non-Abelian gauge geometry 65
  and the Dirac equation 1
  and the electron's magnetic moment 118
  and the Higgs phenomenon 41
  as a vortex 43
  as a vortex line 61
  classical 113
  existence of 47
  experimental observation of 43
  field 48, 49
  from first principles 28
  from the Dirac equation 22
  fundamental magnetizing field 33
  in classical physics 127
  in GWS theory 103
  in magnetization by light 95
  in magneto-optics 129
  in the electromagnetic sector 65
  in the group O(3) 56
  in the Lorentz group 16
  in the vacuum 56, 59, 126, 131
  in type II superconductors 48
  in unified field theory 95
  is a vortex line 60
  magnetic field 100
  magnetic field of light 1
  magnetizing field of light 8
  photon mass and the spin field 41
  physical field 92
  physical field of light 107
  spin field(s) 1, 5, 8, 41, 80, 130
  square root power density fingerprint 138
  stability in free space 49
  vacuum spin fields 79
  vacuum vortex line 61
  vector potential 37
  vortex 50
  with photon mass 56
Bianchi identity 144
Born interpretation 3
Breaking of the vacuum symmetry 131

Canonical
  momentum 114
  quantization 50
  quantization and $B^{(3)}$ 113
CERN experiment 95, 131
Charge quantization condition 69, 77, 79, 93, 99
Circular basis (1), (2), (3) 135, 139
Circular representation
  of 3-D configuration space 111
  of isospin space 136
Classical
  equation of motion of e in A_μ 37
  Hamilton-Jacobi equation 17
  Proca equation 53
  relativistic Hamilton-Jacobi equation 67
Commutator of covariant derivatives 98
Complex
  Klein-Gordon equation 67
  Klein-Gordon field 50

Configuration space in the circular basis (1), (2) and (3)   104
Conjugate
   plane waves   154
   products   31, 34, 66, 81, 85, 113, 123, 130, 139
   products $B^{(1)} \times B^{(2)}$ 126
Connection coefficients   143
Continuity equation   4
Convergent vertex contribution   119
Covariant derivative(s)   62, 102, 135, 143
Creation and annihilation operators   116
Current-density four-vector   3
Curvature tensor   143
Cyclic relations for vector potentials   157
Cyclically symmetric equations for finite photon mass   44

D'Alembert equation   7
D'Alembertian   22
De Broglie's Guiding Theorem   46
Dimensional
   normalization   121
   regularization   117
Dirac
   adjoint spinor   8, 24
   condition   131
   current   120
   four-spinor   8, 18
   matrices   8, 20
   sea   1, 25
Dirac equation(s)   6, 21, 23, 25, 26
   and the SL(2,C) group   15
   $B^{(3)}$ 22
   free particle   19
   of e in A_μ 17, 28,
Divergence of a Feynman graph   117
Dual tensor   82

E(2) group   107
E(2) little group   130
Effective vacuum current   82
Einstein
   equation for rest energy   2
   field equations   134, 144
Electromagnetic
   field massive   114
   mass   58
Electromagnetism   52
   and non-Abelian gauge field   65
   in the vacuum   132
Electron
   magnetic moment and $B^{(3)}$ 118
   momentum operator   35
   moving at the speed of light   133
   neutrino   95
   plasma   93
Electronic cyclotron frequency   38
Energy eigenvalues   25
Equation of charge quantization   74

Euler-Lagrange equation   114
Exchange of virtual photons   134

Faraday induction   83
Fermi-Dirac statistics   5
Fermions   5
Feynman parameter   120
Field tensor $G_{\mu\nu}$ 75
Fine structure constant   118
Flux vortex line   63
Four component spinors   16
Four-spinor(s)   5, 20
Free electron   1
Free photon momentum   76
Fundamental gauge theory   139

Gauge
   field A_μ 3
   field(s)   6, 106
   geometry planar   157
   invariance   51, 131
   transformation   29
Gell-Mann-Nishijima relation   105
General relativity   70, 132
Geometrical theory of gauge fields   68
Geometry of gauge fields   65
Gravitation
   constant   145
   field   70
Group of electrodynamics   85
Gupta-Bleuler condition   158
GWS
   Lagrangian   107
   theory   97

Half integral spin   13
Hamilton-Jacobi equation   125
   relativistic   125
Heisenberg commutators   115
Helicity   9, 22
Helium gas   124
Hermitian transpose   10
Higgs
   boson   47, 56
   field   41
   mechanism   131
   phenomenon   48, 109
HP monopoles   91

$I_0^{1/2}$ dependence   17, 124, 126
   and $B^{(3)}$ 128, 132
   and $B^{(3)}$ from microwave pulses interacting with   134
   of $B^{(3)}$ 124, 130
Indices of the circular basis   97
Infinitesimal rotation generators   14
Instanton solution   91
Interaction at a distance   70
Intrinsic photon mass   59

Index

Intrinsic spin 13
  angular momentum operator 32
  electronic spin 1, 32, 119
Inverse and optical Faraday effects 65
Inverse Faraday effect 47, 56, 126, 138
  semi-classical description of 127
Iso-coordinate system 72
Isospin and gauge symmetry 92
Isospin indices 66
Isospinor 95
Isotopic spin 111

Jacobi identity 71

Klein-Gordon equation 1, 2

Lagrange equation 50
Landé factor 118, 120
Lie algebra 16
Light shifts 126
Light squeezing in quantum optics 121
Little group 107
London equation in the vacuum 54
Longitudinal
  angular momentum of the photon 116
  component $A^{(3)}$ 43
  spin field 41
Lorentz boost 9, 19
  transformation matrix 27
Lorentz group 6, 16

Magnetic
  effects of light 69
  monopoles 49
  plane waves 34
Magnetization 83
  by circularly polarized electromagnetic radiation 123
  by electromagnetic radiation 107
  by light 47, 127
  by microwave pulses 127
  $M^{(3)}$ 125
  of an electron plasma 66, 124
Magnetizing field of light 1
Magneto-optic
  and $B^{(3)}$ 129
  effects 130
Mass parameter 51
Mass term 108
Matter wave 44
Maxwell equations 17
  in matter 7
Metric tensor 144
Microwave
  pulses, magnetization by 127
  radiation 124
  radiation circularly polarized 124

Minimal prescription 28, 34

NAE and general relativity 143
Neutral
  current processes 106
  weak boson 104
Neutrino interaction 104
New quantum theory 2
Noether's Theorem 4
Non-Abelian
  algebra 48
  charge 95
  definition of $F_{\mu\nu}$ 68
  electrodynamics (NAE) 69
  electrodynamics field tensor $G_{\mu\nu}^{(1)}$ 147
  electromagnetic field 69
  field four-tensor $G_{\mu\nu}$ 73
  field theory 66
  gauge fields 92
  gauge geometry 70
  $G_{\mu\nu}^{a}$ 143
  group 65
  Maxwell equations 147
  Maxwell equations in the vacuum 159
  rotation generators 71
  SSB 106
Non-linear field theories 49
Non-zero photon mass 109
Normalized units 20

O(3) group 5, 135
  $B^{(3)}$ in 56
  classical electrodynamics 89
  covariant derivative D_μ in the basis (1), (2), (3) 139
  cyclic relations 87
  electrodynamics 77, 80
  electromagnetic field tensor 135
  field tensor 101
  homogeneous Maxwell equations 82, 144
  IME equations 81
  inhomogeneous Maxwell equation in the vacuum 79
  isospin indices 99
  Maxwell equations in the vacuum 79
  potential model 109
  QED 90
  relation $eA^{(0)} = \hbar\kappa$ 133
  renormalization of, QED 89
  rotation 9, 12
  rotation generator(s) 71, 135
  sector 68
O(3) group gauge
  geometry 132
  geometry rigorous theory of 152
  theory 86
One electron hyperpolarizability 125

O(3) group Non-Abelian
  electrodynamics 85
  electromagnetic sector 66
  gauge theory 138
One electron
  hyperpolarizability 125
  susceptibility 125
  theories 124
Optical
  Aharonov-Bohm 126, 128, 129
  Cotton-Mouton effect 129
  ESR 129
  Faraday and Zeeman effects 126
  Faraday effect 129
  Majorana effect 129
  NMR 128, 129
  Zeeman effect 129
Orbital and intrinsic spinning motion of e in A_μ 132
Origin of fermions 13

Parallel transport in isospace 72
Parity
  operator 18
  violation 96
Particulate and undulatory charge 69
Pauli
  matrices 21
  spinors 11
Phase independent magnetic field $B^{(3)}$ 128
Photomagneton 49
  operator 116
Photon
  exchange of 134
  linear momentum 101
  longitudinal polarization 107
  momentum operator 35
Photon mass 41, 47
  finite 130
  in GWS 107
Physical magnetic fields $B^{(1)}, B^{(2)}$ and $B^{(3)}$ 130
Planar gauge geometry 157
Planck law 133
Plane wave spinors 27
Poincaré group 49
Potential four-vector 51
Power counting argument 90
Primitively divergent graphs 90
Principle of least action 131
  and $B^{(3)}$ 127
Probability current 3
  and density 23
Probability density 7
Proca equation 3, 6, 23, 45, 47, 116, 131
Product group of GWS 95

Quantization condition 132
  charge 74, 147, 159

Quantization of special relativity 2
Quantized field charge 69, 70
Quantum electrodynamics (QED) 113
  $B^{(3)}$ in 113, 118
  the effect of $B^{(3)}$ on renormalizability 116
Quantum mechanical axioms 2
Quantum optics light squeezing in 121

Radio frequency magnetization 93
Radius of the universe 60
Range of electromagnetism 43
Relativistic quantum mechanics 22
Rest frequency of radiation of finite mass 46
Rest wave vector 44
Ricci tensor 144
Riemann-Christoffel curvature tensor 143

Scalar field 2
Scalar particle 2
Schrödinger equation 3
Self-propagating gravitational field 134
Semi-classical theories 129
Semi-commutators 36
SL(2,C) group
  and the Dirac equation 15
  space-time 5
  spinors of 16
Slavnov-Taylor identities 90
Soliton solution 49
Solitons 49
Solutions of Proca's equation 47
Space-time 52
  geometry 8
Special relativity 2
Spin Components 6
Spin trajectory of e in A_μ 126
Spinors 1
Spontaneous symmetry breaking 41, 47, 55, 59
Square root $I_0^{1/2}$ dependence 17
Square root power density dependence 61, 123
SQUID device 128
Standard representation 26, 27
Stern-Gerlach experiment 1
String condition 49
Stringlike magnetic flux 49
SU(2) group 5
  commutators 16
  spinors of 9
  SU(2) ⊗ SU(2) product 18
Symmetry broken Higgs Lagrangian 56
Symmetry broken Lagrangian 62

Theory of gauge field geometry 66

Index

Three degrees of spacelike polarization 41
Tired light 56
Topology of the vacuum 47
Total relativistic energy 20
Trajectory of e in A_μ 131
True derivative 72

Unified field theory 65
Unitary matrices 9
Unitary transformation matrix 10

Vacuum
  eigenstates 55
  expectation value 55
  friction 55
  resistance R 54
  state of electromagnetism 60
Vector bosons 90

Ward identity 90
Weak hypercharge 105
Weak isospin 95, 105
Weinberg angle 97
Weyl equation(s) 7, 21

Yang-Mills formulation 79

Contents of Volume 1

PREFACE		ix
1.	**WAVE AND PARTICLE**	1
	1.1. The Enigma of Wave and Particle: Planck, Einstein, De Broglie	1
	1.2. Symmetry in Classical Electrodynamics, Wavenumber and Linear Momentum	6
	1.3. The Effect of $B^{(3)}$ on the Fundamentals of the Old Quantum Theory	17
2.	**FUNDAMENTAL SYMMETRIES**	21
	2.1. The Seven Discrete Symmetries of Nature	21
	2.2. The $\hat{C}\hat{P}\hat{T}$ Theorem	24
	2.2.1. $\hat{P}$ Conservation	25
	2.2.2. $\hat{T}$ Conservation	26
	2.3. The Concept of Photon and Anti-Photon	26
	2.4. Parity of the Photon and Anti-Photon	28
	2.5. Motion Reversal Symmetry of Photon and Anti-Photon	29
	2.6. The Photon's Charge Conjugation Symmetry, $\hat{C}$	30
	2.7. Scheme for the Photon's Fundamental Symmetries, $\hat{C}$, $\hat{P}$ and $\hat{T}$	30
	2.8. Symmetry of the Pure Imaginary Field $-iE^{(3)}/c$	33
	2.9. Experimental Demonstration of the Existence of the Anti-Photon	35

3. THE ORIGINS OF WAVE MECHANICS — 37

3.1. The Phase as Wave Function — 39
3.2. The Wave Mechanics of a Single Photon in Free Space — 40
3.3. Stationary States of One Photon in Free Space — 42
3.4. Heisenberg Uncertainty and the Single Photon — 46

4. INTER-RELATION OF FIELD EQUATIONS — 55

4.1. Relation Between the Dirac and D' Alembert Wave Equations — 55
4.2. Equations of the Quantum Field Theory of Light — 59
4.3. D'Alembert and Proca Equations — 65

5. TRANSVERSE AND LONGITUDINAL PHOTONS AND FIELDS — 71

5.1. Axial Unit Vectors, Rotation Generators, and Magnetic Fields — 72
5.2. Polar Unit Vectors, Boost Generators, and Electric Fields — 77
5.3. Lie Algebra of Electric and Magnetic Fields in the Lorentz Group, Isomorphism — 80
5.4. The Eigenvalues of the Massless and Massive Photons: Vector Spherical Harmonics and Irreducible Representations of Longitudinal Fields — 82

6. CREATION AND ANNIHILATION OF PHOTONS — 89

6.1. The Meaning of Photon Creation and Annihilation — 89
6.2. Quantum Classical Equivalence — 91
6.3. Longitudinal and Time-Like Photon Operators; Bilinear (Photon Number) Operators — 93
6.4. Light Squeezing and the Photomagneton $\hat{B}^{(3)}$ — 98

Contents

7. EXPERIMENTAL EVIDENCE FOR $\hat{B}^{(3)}$ — 103

    7.1.   The Inverse Faraday Effect, Magnetization by Light — 105
    7.2.   Optical NMR, First Order Effect of $B^{(3)}$ — 108
    7.3.   The Optical Faraday Effect (OFE) — 111
    7.4.   Survey of Data — 115

8. THE CONCEPT OF PHOTON MASS — 117

    8.1.   The Problem — 117
    8.2.   Brief Review of Experimental Evidence Compatible with $m_0 \neq 0$ and $B^{(3)} \neq 0$ — 120
    8.3.   The Proca Equation — 123
            8.3.1. Lorentz Invariance of the Maxwell Equations in Free Space — 125
            8.3.2. Covariance of the Proca Equation in Free Space — 126
    8.4.   Analogy between Photon Mass and Effective Current — 126
    8.5.   General Solutions of the Proca Equation — 129

9. AHARONOV-BOHM EFFECTS — 133

    9.1.   The Original Theory of Aharonov and Bohm — 134
    9.2.   The Effect of the Cyclic Algebra (25) — 136
    9.3.   The Optical Aharonov-Bohm Effect — 136
    9.4.   The Physical A_μ and Finite Photon Mass — 137
    9.5.   If the OAB Is Not Observed — 139
    9.6.   Non-Linearity of Photon Spin in Free Space — 142

10. MODIFICATIONS OF LAGRANGIAN FIELD THEORY — 147

    10.1.   Novel Gauge Fixing Term — 150
    10.2.   Quantization of the Electromagnetic Field — 152
    10.3   A Potential Model for $B^{(3)}$ — 158

11. PSEUDO FOUR-VECTOR REPRESENTATIONS OF ELECTRIC AND MAGNETIC FIELDS — 161

    11.1. Relation between the Minkowski and Lorentz Forces — 163
    11.2. Dual Pseudo Four-Vector of $F_{\rho\sigma}$ in Free Space — 165
    11.3. Link between V_μ, B_μ and E_μ — 166
    11.4. Some Properties of E_μ and B_μ in Free Space — 167
    11.5. Consequences for the Fundamental Theory of Free Space Electromagnetism — 169
    11.6. Consequences for the Theory of Finite Photon Mass — 169

12. DERIVATION OF $B^{(3)}$ FROM THE RELATIVISTIC HAMILTON-JACOBI EQUATION OF e IN A_μ — 171

    12.1. Action and the Hamilton-Jacobi Equation of Motion — 171
    12.2. Solution of the Relativistic Hamilton-Jacobi Equation (369) — 174
    12.3. The Orbital Angular Momentum of the Electron in the Field — 178
    12.4. Limiting Forms of Equation (405) — 179
    12.5. Discussion — 180

APPENDICES

A. Invariance and Duality in the Circular Basis — 185

B. Angular Momentum in Special Relativity — 189

C. Standard Expressions for the Electromagnetic Field in Free Space, with Longitudinal Components — 193

D. The Lorentz Force Due to $F_{\mu\nu}^{(3)}$; and $T_{\mu\nu}^{(3)}$ — 199

REFERENCES — 205

INDEX — 213

Fundamental Theories of Physics

Series Editor: Alwyn van der Merwe, *University of Denver, USA*

1. M. Sachs: *General Relativity and Matter.* A Spinor Field Theory from Fermis to Light-Years. With a Foreword by C. Kilmister. 1982 ISBN 90-277-1381-2
2. G.H. Duffey: *A Development of Quantum Mechanics.* Based on Symmetry Considerations. 1985 ISBN 90-277-1587-4
3. S. Diner, D. Fargue, G. Lochak and F. Selleri (eds.): *The Wave-Particle Dualism.* A Tribute to Louis de Broglie on his 90th Birthday. 1984 ISBN 90-277-1664-1
4. E. Prugovečki: *Stochastic Quantum Mechanics and Quantum Spacetime.* A Consistent Unification of Relativity and Quantum Theory based on Stochastic Spaces. 1984; 2nd printing 1986 ISBN 90-277-1617-X
5. D. Hestenes and G. Sobczyk: *Clifford Algebra to Geometric Calculus.* A Unified Language for Mathematics and Physics. 1984
   ISBN 90-277-1673-0; Pb (1987) 90-277-2561-6
6. P. Exner: *Open Quantum Systems and Feynman Integrals.* 1985 ISBN 90-277-1678-1
7. L. Mayants: *The Enigma of Probability and Physics.* 1984 ISBN 90-277-1674-9
8. E. Tocaci: *Relativistic Mechanics, Time and Inertia.* Translated from Romanian. Edited and with a Foreword by C.W. Kilmister. 1985 ISBN 90-277-1769-9
9. B. Bertotti, F. de Felice and A. Pascolini (eds.): *General Relativity and Gravitation.* Proceedings of the 10th International Conference (Padova, Italy, 1983). 1984
   ISBN 90-277-1819-9
10. G. Tarozzi and A. van der Merwe (eds.): *Open Questions in Quantum Physics.* 1985
    ISBN 90-277-1853-9
11. J.V. Narlikar and T. Padmanabhan: *Gravity, Gauge Theories and Quantum Cosmology.* 1986 ISBN 90-277-1948-9
12. G.S. Asanov: *Finsler Geometry, Relativity and Gauge Theories.* 1985
    ISBN 90-277-1960-8
13. K. Namsrai: *Nonlocal Quantum Field Theory and Stochastic Quantum Mechanics.* 1986 ISBN 90-277-2001-0
14. C. Ray Smith and W.T. Grandy, Jr. (eds.): *Maximum-Entropy and Bayesian Methods in Inverse Problems.* Proceedings of the 1st and 2nd International Workshop (Laramie, Wyoming, USA). 1985 ISBN 90-277-2074-6
15. D. Hestenes: *New Foundations for Classical Mechanics.* 1986
    ISBN 90-277-2090-8; Pb (1987) 90-277-2526-8
16. S.J. Prokhovnik: *Light in Einstein's Universe.* The Role of Energy in Cosmology and Relativity. 1985 ISBN 90-277-2093-2
17. Y.S. Kim and M.E. Noz: *Theory and Applications of the Poincaré Group.* 1986
    ISBN 90-277-2141-6
18. M. Sachs: *Quantum Mechanics from General Relativity.* An Approximation for a Theory of Inertia. 1986 ISBN 90-277-2247-1
19. W.T. Grandy, Jr.: *Foundations of Statistical Mechanics.*
    Vol. I: *Equilibrium Theory.* 1987 ISBN 90-277-2489-X
20. H.-H von Borzeszkowski and H.-J. Treder: *The Meaning of Quantum Gravity.* 1988
    ISBN 90-277-2518-7
21. C. Ray Smith and G.J. Erickson (eds.): *Maximum-Entropy and Bayesian Spectral Analysis and Estimation Problems.* Proceedings of the 3rd International Workshop (Laramie, Wyoming, USA, 1983). 1987 ISBN 90-277-2579-9

Fundamental Theories of Physics

22. A.O. Barut and A. van der Merwe (eds.): *Selected Scientific Papers of Alfred Landé.* [*1888-1975*]. 1988 ISBN 90-277-2594-2
23. W.T. Grandy, Jr.: *Foundations of Statistical Mechanics.*
   Vol. II: *Nonequilibrium Phenomena.* 1988 ISBN 90-277-2649-3
24. E.I. Bitsakis and C.A. Nicolaides (eds.): *The Concept of Probability.* Proceedings of the Delphi Conference (Delphi, Greece, 1987). 1989 ISBN 90-277-2679-5
25. A. van der Merwe, F. Selleri and G. Tarozzi (eds.): *Microphysical Reality and Quantum Formalism, Vol. 1.* Proceedings of the International Conference (Urbino, Italy, 1985). 1988 ISBN 90-277-2683-3
26. A. van der Merwe, F. Selleri and G. Tarozzi (eds.): *Microphysical Reality and Quantum Formalism, Vol. 2.* Proceedings of the International Conference (Urbino, Italy, 1985). 1988 ISBN 90-277-2684-1
27. I.D. Novikov and V.P. Frolov: *Physics of Black Holes.* 1989 ISBN 90-277-2685-X
28. G. Tarozzi and A. van der Merwe (eds.): *The Nature of Quantum Paradoxes.* Italian Studies in the Foundations and Philosophy of Modern Physics. 1988
   ISBN 90-277-2703-1
29. B.R. Iyer, N. Mukunda and C.V. Vishveshwara (eds.): *Gravitation, Gauge Theories and the Early Universe.* 1989 ISBN 90-277-2710-4
30. H. Mark and L. Wood (eds.): *Energy in Physics, War and Peace.* A Festschrift celebrating Edward Teller's 80th Birthday. 1988 ISBN 90-277-2775-9
31. G.J. Erickson and C.R. Smith (eds.): *Maximum-Entropy and Bayesian Methods in Science and Engineering.*
   Vol. I: *Foundations.* 1988 ISBN 90-277-2793-7
32. G.J. Erickson and C.R. Smith (eds.): *Maximum-Entropy and Bayesian Methods in Science and Engineering.*
   Vol. II: *Applications.* 1988 ISBN 90-277-2794-5
33. M.E. Noz and Y.S. Kim (eds.): *Special Relativity and Quantum Theory.* A Collection of Papers on the Poincaré Group. 1988 ISBN 90-277-2799-6
34. I.Yu. Kobzarev and Yu.I. Manin: *Elementary Particles. Mathematics, Physics and Philosophy.* 1989 ISBN 0-7923-0098-X
35. F. Selleri: *Quantum Paradoxes and Physical Reality.* 1990 ISBN 0-7923-0253-2
36. J. Skilling (ed.): *Maximum-Entropy and Bayesian Methods.* Proceedings of the 8th International Workshop (Cambridge, UK, 1988). 1989 ISBN 0-7923-0224-9
37. M. Kafatos (ed.): *Bell's Theorem, Quantum Theory and Conceptions of the Universe.* 1989 ISBN 0-7923-0496-9
38. Yu.A. Izyumov and V.N. Syromyatnikov: *Phase Transitions and Crystal Symmetry.* 1990 ISBN 0-7923-0542-6
39. P.F. Fougère (ed.): *Maximum-Entropy and Bayesian Methods.* Proceedings of the 9th International Workshop (Dartmouth, Massachusetts, USA, 1989). 1990
   ISBN 0-7923-0928-6
40. L. de Broglie: *Heisenberg's Uncertainties and the Probabilistic Interpretation of Wave Mechanics.* With Critical Notes of the Author. 1990 ISBN 0-7923-0929-4
41. W.T. Grandy, Jr.: *Relativistic Quantum Mechanics of Leptons and Fields.* 1991
   ISBN 0-7923-1049-7
42. Yu.L. Klimontovich: *Turbulent Motion and the Structure of Chaos.* A New Approach to the Statistical Theory of Open Systems. 1991 ISBN 0-7923-1114-0

Fundamental Theories of Physics

43. W.T. Grandy, Jr. and L.H. Schick (eds.): *Maximum-Entropy and Bayesian Methods.* Proceedings of the 10th International Workshop (Laramie, Wyoming, USA, 1990). 1991 ISBN 0-7923-1140-X
44. P.Pták and S. Pulmannová: *Orthomodular Structures as Quantum Logics.* Intrinsic Properties, State Space and Probabilistic Topics. 1991 ISBN 0-7923-1207-4
45. D. Hestenes and A. Weingartshofer (eds.): *The Electron.* New Theory and Experiment. 1991 ISBN 0-7923-1356-9
46. P.P.J.M. Schram: *Kinetic Theory of Gases and Plasmas.* 1991 ISBN 0-7923-1392-5
47. A. Micali, R. Boudet and J. Helmstetter (eds.): *Clifford Algebras and their Applications in Mathematical Physics.* 1992 ISBN 0-7923-1623-1
48. E. Prugovečki: *Quantum Geometry.* A Framework for Quantum General Relativity. 1992 ISBN 0-7923-1640-1
49. M.H. Mac Gregor: *The Enigmatic Electron.* 1992 ISBN 0-7923-1982-6
50. C.R. Smith, G.J. Erickson and P.O. Neudorfer (eds.): *Maximum Entropy and Bayesian Methods.* Proceedings of the 11th International Workshop (Seattle, 1991). 1993 ISBN 0-7923-2031-X
51. D.J. Hoekzema: *The Quantum Labyrinth.* 1993 ISBN 0-7923-2066-2
52. Z. Oziewicz, B. Jancewicz and A. Borowiec (eds.): *Spinors, Twistors, Clifford Algebras and Quantum Deformations.* Proceedings of the Second Max Born Symposium (Wrocław, Poland, 1992). 1993 ISBN 0-7923-2251-7
53. A. Mohammad-Djafari and G. Demoment (eds.): *Maximum Entropy and Bayesian Methods.* Proceedings of the 12th International Workshop (Paris, France, 1992). 1993 ISBN 0-7923-2280-0
54. M. Riesz: *Clifford Numbers and Spinors* with Riesz' Private Lectures to E. Folke Bolinder and a Historical Review by Pertti Lounesto. E.F. Bolinder and P. Lounesto (eds.). 1993 ISBN 0-7923-2299-1
55. F. Brackx, R. Delanghe and H. Serras (eds.): *Clifford Algebras and their Applications in Mathematical Physics.* Proceedings of the Third Conference (Deinze, 1993) 1993 ISBN 0-7923-2347-5
56. J.R. Fanchi: *Parametrized Relativistic Quantum Theory.* 1993 ISBN 0-7923-2376-9
57. A. Peres: *Quantum Theory: Concepts and Methods.* 1993 ISBN 0-7923-2549-4
58. P.L. Antonelli, R.S. Ingarden and M. Matsumoto: *The Theory of Sprays and Finsler Spaces with Applications in Physics and Biology.* 1993 ISBN 0-7923-2577-X
59. R. Miron and M. Anastasiei: *The Geometry of Lagrange Spaces: Theory and Applications.* 1994 ISBN 0-7923-2591-5
60. G. Adomian: *Solving Frontier Problems of Physics: The Decomposition Method.* 1994 ISBN 0-7923-2644-X
61. B.S. Kerner and V.V. Osipov: *Autosolitons.* A New Approach to Problems of Self-Organization and Turbulence. 1994 ISBN 0-7923-2816-7
62. A. Heidbreder (ed.): *Maximum Entropy and Bayesian Methods.* Proceedings of the 13th International Workshop (Santa Barbara, USA, 1993) 1995 ISBN 0-7923-2851-5
63. J. Peřina, Z. Hradil and B. Jurčo: *Quantum Optics and Fundamentals of Physics.* 1994 ISBN 0-7923-3000-5

Fundamental Theories of Physics

64. M. Evans and J.-P. Vigier: *The Enigmatic Photon*. Volume 1: The Field $B^{(3)}$. 1994
   ISBN 0-7923-3049-8
65. C.K. Raju: *Time: Towards a Constistent Theory*. 1994   ISBN 0-7923-3103-6
66. A.K.T. Assis: *Weber's Electrodynamics*. 1994   ISBN 0-7923-3137-0
67. Yu. L. Klimontovich: *Statistical Theory of Open Systems*. Volume 1: A Unified Approach to Kinetic Description of Processes in Active Systems. 1995
   ISBN 0-7923-3199-0; Pb: ISBN 0-7923-3242-3
68. M. Evans and J.-P. Vigier: *The Enigmatic Photon*. Volume 2: Non-Abelian Electrodynamics. 1995   ISBN 0-7923-3288-1

KLUWER ACADEMIC PUBLISHERS – DORDRECHT / BOSTON / LONDON